Astronomers Anonymous

Steve Ringwood

Astronomers Anonymous

Getting Help with the Puzzles and Pitfalls
of Practical Astronomy

 Springer

Steve Ringwood
School Green Lane 31
CM16 6EH Epping
North Weald
UK

ISBN 978-1-4419-5816-7   e-ISBN 978-1-4419-5817-4
DOI 10.1007/978-1-4419-5817-4
Springer New York Dordrecht Heidelberg London

Library of Congress Control Number: 2010928644

Printed on acid-free paper

Springer is part of Springer Science+Business Media (www.springer.com)

This book is warmly dedicated to amateur astronomers who have stood drenched beneath driving rain during an eclipse, dropped an expensive eyepiece to an untimely doom, watched a precisely timed lonely cloud snuff out an essential observation, or suffered any of the maladies (self inflicted or otherwise) that befall those who sacrifice all to glimpse the universe. Yes, that's every single one of us. We are not alone.

Preface

During more than four decades of involvement in amateur astronomy, I have enjoyed the privilege of rubbing shoulders with numberless amateur and professional astronomers. In so doing I have encountered at first, second, and third hand many of the joys and pitfalls that sky watchers can experience in pursuit of the universe's wonders. I have often howled at tall tales that would not disgrace a pirate's tavern. Many of these astounding stories have become the kernels of my Dear Steve column items.

Learning how to operate the technology for observing and imaging the universe is work enough for any aspiring astronomer; however, many have problems of their own making. Not only do they share these troubles with other astronomers, they are on the receiving end of colleagues and friends doing the same. With all these agonized communications flying about, it is hard to understand how anyone gets any real work done!

For the amusement of my peers I have long fondly parodied these imagined literary exchanges. These fantasy "Agony Aunt" questions began appearing in the pages of the Loughton Astronomical Society's monthly (and Christmas Special) journals about 30 years ago, in the guise of *The astronomer's problem page*. This was by the kind indulgence of the then editor, namely myself.

Happily, even when the magazine of the LAS evolved into something much better, under the tender and loving care of those who came after me, these problem letters were still in demand and even now occasionally appear.

In this omnibus collection of letters, herein addressed to my "virtual" self, I have excluded many of the original letters because they were too scandalous, libelous, or inciteful (and even insightful). Some barely made it under the wire only by dint of subtle amendment to protect the guilty! Inspired by the present opportunity to reveal further mental tortures of my contemporaries, a large number of new correspondences make their first appearance here.

Many of the featured problems are heavily disguised authentic ones experienced by real amateur astronomers, though it need hardly be said many are not; especially (he adds hurriedly) those of a deviant or diseased nature. In fact it disturbs me a little to find an inordinate number of those that survived the initial cull still embody both of these! After all, their conception has been forged by my own experiences. Quite a few are barely fictionalized versions of real problems related by professional astronomers. I leave it to you to guess which these are.

This work serves three masters. First, it can be read with knowing amusement by experienced amateur astronomers, who no doubt will see in these pages an alarming but reassuring reflection of themselves. Second, it can act as a warning to the unwary novice, teetering momentarily in the balance between a vague, barely indulged astronomical curiosity and reckless abandonment into scientific enthusiasm. Thirdly, it offers valuable insight for those interested outsiders who want to know what on Earth makes these strange people tick.

Above all, this book's purpose is to entertain. I have liberally laced it with supplementary factual information that may hopefully prove informative, thoughtful, and even useful. So turn the page (with trepidation) and hopefully enjoy. But warn your friends. Spare them the pain.

North Weald, England   Steve Ringwood
Fall 2009

Acknowledgements

I must express my huge appreciation to two people who corrupted their sensibilities by reading through the more than 130 original correspondences to weed out those that were either unsuitable or simply inadequate.

Fellow astronomer Martin Peston risked arrest by reading them while traveling on public transport. Regardless of the commuter density, his frequent maniacal laughter ensured that once he started reading he always had a seat – a choice of several, in fact.

My wife Gillian, a keen astronomer in her own right, has nobly suffered my humor from the first, so ploughed through this task with typical patient resignation and compassion. I am also grateful for her sterling work on the final stages of the manuscript, pointing out many errors, ambiguities, and omissions. She having just earned her Open University degree, I was forced to take notice.

My dad (okay, that's three!) unwisely consented to read through the manuscript in order to confirm that most people will know what the hell I am going on about. He passed it, but I think he may be putting me up for adoption.

So far as some of the factual material goes, I must recognize the unwitting contribution of two people.

Firstly, I give a nod of appreciation to a professional astronomer at the South African Astronomical Observatory. His peripatetic discovery and publication of odd stories from astronomy's professional archives informs some of the items I allude to. I need hardly add that he is a prime example of that scientific principle known as nominative determinism – being a man who uses telescopes, called I. Glass!

The second silent contributor is the greatly missed historian of science Colin Ronan, who in the years I knew him was a constant fund of wonderfully humorous and scandalous stories from the scientific world. I only wish that I could recall more of them.

Last, but not least, this book would not have seen the dark of night had it not been for John Watson (Astrobooks), who wanted to find out if Springer had a sense of humor. Maury Solomon of Springer, too, deserves praise for mirthfully falling off her stool to confirm that they have.

Pre-release Reviews . . . .

Aristarchus of Samos – Unlike the Sun, this will be the centre of no-one's Universe.

Eratosthenes – This author writes total rubbish. The Earth is too small for the both of us.

Ptolemy – An epicyclic piece of work if ever I saw one. I am completely indeferent.

Tycho Brahe – I would cut off my nose rather than read this drivel.

Nicolaus Copernicus – His central arguments are completely circular. I would never be so silly.

Galileo Galilei – You'd think that with my 'scopes I'd have seen this book coming!

Edmond Halley – If he writes another book within 76 years it'll be 80 years too soon

Isaac Newton – This upstart Ringwood does not give his subject sufficient gravity.

John Dollond – Completely lacking in color.

Charles Messier – I have placed this object at once in my catalogue - to avoid its confusion with a real book.

Giovanni Cassini – My opinion is completely divided on its merits.

William Herschel – I discovered a planet, then discovered this book. What a letdown.

Johannes Bode – He's just playing with numbers I tell you. They mean nothing

Joseph Fraunhofer – There are a few lines here of interest, but little else.

Percival Lowell – He may think his wise cracks are funny, but look what happened when I described mine!

Lord Kelvin – Not a hot product, to any degree.

Heinrich Olbers – If the book is so great, why isn't it selling? Definitely a paradox here.

Edwin Hubble – The further away I get from this book, the faster I can forget it.

Edwin Schrodinger – I put this book in a cat box. Now I can't decide if it's readable or not.

Albert Einstein – This book is complete fudge, a factor I would never use in my work.

Contents

About the Author

Steve Ringwood was given his first astronomy book (*The Golden Book of Astronomy*) when he was nine, and, like many others of his generation, the Apollo years cemented his lifelong fascination with space. At 15 he bought his first telescope, a Japanese 40 mm refractor. That first incredible view of a first quarter Moon had him hooked, and thus began the series of instrument acquisitions of steadily growing apertures that continues to this day.

As a teen he joined the British Astronomical Association (BAA) and shortly afterwards joined his local astronomy group (Loughton Astronomical Society), which, at the time, was building its own observatory to house a 16″ Cassegrain reflector. He later served many years on its committee – several of them as its chairman.

During his 40 years of astronomical activity Steve has encountered at first, second, and third hand many of the joys, pitfalls, and anxieties that astronomers of all shades face. His interest in astronomy and spaceflight has taken him all over the world – from the winter bleakness of northern Russia to the heat of northwestern Africa and the tropical wonders of Hawaii. Telescopically, his preferred targets are within the Solar System: the planets, the Sun, and the Moon. He also has a keen interest in the history of astronomy.

In addition to at one time editing his local group's monthly journal, he has written for both US and UK astronomy magazines. Together with occasional features and reviews, he currently produces a monthly product column for *Astronomy Now*. There have also been occasional "blink and you'll miss it" interactions with TV and radio.

Despite a long career in IT, his passion is astronomy and palaeontology (fossil collecting). Elected a Fellow of the Royal Astronomical Society in 1984, he has written papers for the journals of both the BAA and the Royal Astronomical Society.

With his wife and young son, he lives in a house chosen partly for its distance from sodium glow at night – a remoteness sadly decreasing with the passing years.

Astronomers Anonymous

CHAPTER 1
Some Background

It is often noted (by me for a start) that astronomy is one of the few sciences in which the amateur can perform not simply a useful role but, in many of its fields, play an essential part in contributing time consuming, laborious (unpaid!), yet invaluable observations to professional research. Part of this usefulness derives from the amateurs' outnumbering of professionals by several orders of magnitude.

The obvious result of this disparity is that there will be a much larger number of unpaid observers suffering for their science than those who *are* paid. This seems unfair. But it does mean that this book will be most helpful to those who do this strange thing simply for the love of doing it. There, does that make you feel better?

Some Dear Steve letters appear to focus on medical problems. Whether this reflects personal experience or a cunning extrapolation from orthodox agony columns I do not know.

In the real world, the health problems of astronomers reassuringly seem to reflect those of the general population. However, I would say that this is not entirely what might be expected, given the invigorating environmental extremes that those of our ilk are exposed to.

Because of the advantageous observational conditions (good "seeing") of those long winter nights of steady chilled air, observers think nothing of teasing hypothermia and frostbite for hours — warmed and distracted by the distant icy fires in the sky. Thawing out is generally achieved through solar eclipse trips to scorching arid countries where not even the flies venture out into the midday Sun.

Such "training" certainly toughens astronomers against many purely physical ailments, but perhaps not those of the mind. It is therefore no surprise that problems of the cerebral kind are also featured here, their manifestations possibly the origin for the view askance with which more sensible mortals view astronomers. What other opinion can be expected about those who willingly leave the flaming hearth of a family gathering to venture out to the heat-leaching deep freeze of a January night, merely to gaze upon indistinct specks of light whose images are better seen in any color supplement. For myself, I have no problem in being thought odd. I crave it. I bathe in its eccentric glory. If this opinion wavered, it would alarm me to the possibility of having succumbed unawares to the

S. Ringwood, *Astronomers Anonymous*, DOI 10.1007/978-1-4419-5817-4_1,
© Springer Science+Business Media, LLC 2010

invidious power of common sense. It is at this point that my wife, checking through the final manuscript, wrote boldly in the margin "YOU ARE ODD!" I am happy to discover that I am safe, at least for the moment.

In the world of telescope and accessory acquisition, amateur astronomy fields devotees just as crazy for the latest consumer technology as any sound system expert or computer buff. This engenders an arms race between astronomers as vicious as those between antagonistic super powers.

The various telescopic "arms dealers" are as delighted to feed this competitive frenzy as their military counterparts. Indeed, the correlation between telescopic and martial technologies is uncomfortably close. New developments in ground-to-air missile guidance systems translate freely into the auto guide/tracking systems of telescope mounts; surveillance satellites drive improvements in optics and electronic image sensors. Indeed, many amateur telescopes now employ GPS satellite systems that allegedly are entirely financed and controlled by the military.

Some may feel that the psychological stresses of "gadget fever" portrayed in these pages are somewhat overstated for comedic purpose. If only that were so.

Astronomy, as a subject, is a pursuit in which you can either stay comfortably in the shallows dipping your toe or be dragged by strong currents out to the profoundest depths, with sufficient enthusiasm to control every aspect of daily life. Devotees will of course laud the latter as a justifiable contribution to the sum of all human knowledge. Yet the sky is a harsh mistress. Excessive attention away from the dull but necessary routines of normal familial life creates conflicts for which the enlightenment of humankind is generally seen as insufficient excuse. Not one of us, if we are honest, can say we have never shunted finance or time into astronomy's ever gaping maw that has not been desperately required elsewhere. Therefore, colliding priorities also have a rightful place here.

Each problem contained herein is represented in *italic* type for my imaginary correspondent, with Dear Steve's helpful reply in **bold**. In this book, I have bracketed each correspondence with a gentle commentary in CAPS with addenda in normal type, serving as a precautionary note, genuine piece of information, or advice. After all, let's not waste the opportunity.

Some of the addenda illustrate that professional astronomers have problems just as vexing as their amateur counterparts. I include these in the interest of fairness and to illustrate that those who get paid to do astronomy are in essence the same as those who do not. We are all kids with toys; it's just that those of the professional are a little bigger and more expensive – on the whole.

So that desperate readers may access their own problems quickly I have (very loosely) gathered these letters into seven categories; however, the unafflicted reader is at liberty to dive in recklessly at any point or plod through from the start with dogged determination. I can assure each reader that they will find their own problem here, eventually. Those capable of that lamentable ability called speed-reading are quite at liberty to get the whole messy business over and done with as quickly as possible.

As with all agony columns, similar problems arise again and again with sad inevitability. Yet the solutions proffered by my pseudo-self are inventively varied. Enjoy them all.

We have now arrived at that point in the introduction where traditionally the author, symbolically draped in sackcloth and ashes, takes the blame for all remaining typographic errors, factual errors, irrelevant references, irreverent references, mal-evolent assertions, or unseemly perversions. These are certainly not due to me! I blame everyone else who has been involved.

CHAPTER 2
Fundamentals

Astronomy can justifiably be said to be the science of everything. Its reach encompasses the macrocosm of far-flung space to the microcosm of nuclear physics. Its field of stellar evolution underpins the creation of the elements and even the emergence of life itself. Its dramatic beauty inspires civilizations to great literature, art, and learning.

The unfathomable vastness of this subject matter provides no trammels within which these correspondences must run. Yet there are some problem letters, surprised as I am to find them, that are teetering on the brink of normal astronomical queries. I apologize. This, I am sure, is due to some clumsy oversight on my part.

I do not doubt that real astronomers receive the foregoing types of letters all the time. Perhaps even word for word. In a theoretical

S. Ringwood, *Astronomers Anonymous*, DOI 10.1007/978-1-4419-5817-4_2,
© Springer Science+Business Media, LLC 2010

multiverse of eventualities, it can be proposed that the following exchanges will occur, somewhere, somewhen. I wouldn't want to go there.

(OF THE BIRDS AND THE ER...UM....)

Dear Steve,

I am a very worried twelve year old. When I was seven, I asked my daddy where I came from. He said that I was a little seed that came from the sky and landed in mommy's tummy. I have now heard that some bits of meteorite have been thought to contain life from Mars. Does this mean I am a Martian? (Photo enclosed.)

Sally
Roswell, New Mexico

Dear Sally,

Don't panic! Many fathers unfortunately confuse their children by telling them little fibs about their origin. This is because they try to delay an embarrassing truth until you are better able to understand what really happens. I would suggest you urgently ask him again, since by your obvious curiosity you demonstrate that now is the time for all this messy business to be explained to you. As reassurance in the meantime, might I pass comment on your photo? You look fairly normal, although the surfeit of warty skin and hairy tentacles could indeed be a concern. Please write to me again and let me know what your father says.

It is fascinating to note that blithe acceptance of life elsewhere has risen and fallen quite drastically throughout history. Certainly, there is nothing modern about the public's appetite for news of extraterrestrials. Almost 200 years ago, the world was rocking from the series of revelations that were later to be known as "The Great Lunar Hoax".

In the early 1830s, John Herschel took his attractive young wife, children, and an 18-inch reflector to South Africa to survey the stars of the southern hemisphere. Great things were expected of him. He was, after all, the son of planet discoverer William Herschel (Uranus, 1781), and he would be the first to study those skies with such a large instrument. The world waited with bated breath. They did not have to wait long.

In the August of 1835, an article appeared in the *New York Sun* which, via an alleged intermediary by the name of Dr. Andrew Grant, reported on John Herschel's fantastic discoveries. Using magnifications in the thousands, strange beasts were being observed upon the Moon. These included goats, bison, unicorns, and most amazingly of all, bipedal humanoid bats (named Vespertilio-homo). All these were apparently enjoying an equitable and civilized environment full of lush vegetation, waterfalls, and oceans.

Over the following 5 days, the *New York Sun* related ever wilder observations – all based apparently on the incredible abilities of John Herschel's amazing telescope – although the telescope talked of in these articles was far greater and more powerful than the actual instrument in South Africa. The circulation of the *New York Sun*, which had long been languishing in the doldrums of just a few thousand readers, rocketed by the end of the series of stories to over nineteen thousand.

It is almost certain that the real author of these stories was the Cambridge-educated reporter Richard Adams Locke, although this was never proved. Apart from the opportunistic desire to push up circulation, he may well have been lampooning the Rev. Thomas Dick, a Scottish science teacher and astronomy writer who had recently published "The Christian Philosopher." Dick had claimed that the Solar System was stuffed with other-worldly inhabitants – totaling over 21 trillion; why, Mars alone boasted 4.2 billion.

Although initially amused, Herschel himself found the whole issue rather irksome, since for a long time afterwards many never realized the whole thing was a hoax. His wife, corresponding from South Africa to English friends and relatives at the time of the hoax, talks of his increasing irritability at the story.

John Herschel, writing in January 1837 to his famous aunt Caroline Lucretia Herschel complains, "I have been pestered from all quarters with that ridiculous hoax about the Moon – in English, French, Italian, and German!"

With the recent plethora of extrasolar planetary discoveries (at the end of 2009 ∼400) the idea of alien life is fruiting once more. The number of newly discovered planets is set to accelerate owing to the launch (March 2009) of

Fig 2.1 My own graze with the Herschels. This battered but treasured print records my 1982 encounter with Caroline Herschel. No, not William's comet- discovering sister (I am not THAT old) but with his great-great-granddaughter [Image by the author]

NASA's planet-discovering Kepler mission. Thanks to new techniques for the analysis of their atmospheres, it can only be a matter of time before such presences will be deduced.

(BE MINDFUL THAT ASTRONOMY CAN BE COMPETITIVE, AND THE CANON OF ASTRONOMICAL LITERATURE IS VAST AND HELPFUL.)

Dear Steve,

What measures would you recommend for winning inter-society astronomy quizzes?

Mark
Birmingham, United Kingdom

Dear Mark,

Purchase these excellent books. Read them avidly, then throw them away.
Steve Ringwood's Astronomical Dictionary, by Steve Ringwood $6.99.
Astronomy for Short People, by Steve Ringwood (Thick volume, so multi-purpose) $20.95.
Amateur Astronomy and the Anorak: A Sartorial History, Steve Ringwood $45.
Gocarting for the Amateur Astronomer, by Steve Ringwood $1.50.
Optics; Their Role in My Success, by Patrick Moore $20.50.
Queen; My Role in Their Success, by Patrick Moore $20.50.
Love Life of the Amateur Astronomer, by Steve Ringwood (Slim volume, so only 50 cents.)
How to Win at Quizzes Without Actually Cheating, by Steve Ringwood $5,000.

The allusion to the rock group Queen honors the encouragement given to Queen's Brian May by noted astronomer Sir Patrick Moore, to return to his astronomical studies following an extended break to..er..indulge in musical leanings.

Inter-Astro Society quizzes are fun and a great way to meet other enthusiasts. I have found that very few end up in fist fights. The best way of winning these quizzes is to pick the right team – and those fastest on the buzzer may be just as important as those with the knowledge. Those with the knowledge tend to be the best read. It is therefore worth remembering that basic astronomy books are the books quizmasters use to build quizzes.

(SINCE ANCIENT TIMES ASTRONOMY AND ITS GRUBBIER COUSIN, ASTROLOGY, HAVE EXISTED IN VARIOUS DEGREES OF DISCOMFORT WITH EACH OTHER. THIS IS EXACERBATED BY THE OFT-NOTED OBSERVATION THAT THERE ARE MORE POOR ASTRONOMERS THAN POOR ASTROLOGERS.)

Dear Steve,

I keep getting requests from friends and relatives to do their horoscopes. Whether this is in earnest or fun it is unclear, but they do offer to pay for my services. Being a conscientious astronomer I have so far declined their entreaties. Lately however, my funds have been rather on the thin side so, against my better judgement, I am now tempted to practice this arcane art. But I cannot help feeling that my intellectual integrity would be destroyed forever if I did so. Nevertheless the smell of desperately needed money may prove too much for me. Should I continue to desist, or rake in the loot?

Dan
Little Rock, Arkansas

Dear Dan,

Intellectual integrity is not everything! Believe me! I, too, have suffered overtures, particularly from daily papers I would not wish my name to be associated with. After careful consideration I feel that the best solution in your circumstances is take their money but discourage further pestering. This may seem a difficult achievement simultaneously, but bear with me. What you do is to appear eager to do their horoscope. During the 'process' make sure you say a lot of fancy astronomical words, like conjuncture, quadrature, ephemeral, heliacal, etc. (All these terms and more can be found in the excellent *Astronomical Dictionary for Tricksters and Charlatans, $6.99,* out now in most bookstores.) After the 'consultation,' charge them an absolute fortune. They will be honor-bound to pay up but will refrain from repeating the financially bruising experience. You could also disclose that your analysis reveals them to be a nasty piece of work and totally inadequate in every sense! A threat to make this public may even encourage further remuneration.

Sadly, successful astrologers do get paid more than their astronomical counterparts. The reason for this is simply that there is more demand for astrological material. You need only visit your local bookstore to compare the long heaving multiple level shelves of "new age" material to those of the sciences, the latter normally found crushed into a token cursory minor ledge. Better education, particularly in the sciences, is the only solution. It is worth noting that amateur astronomers are happy to be in the front line of this effort. Nevertheless, always bear in mind that until recent times, astrology was astronomy's paymaster.

The relative poverty of astronomers became particularly ironic in 2004, when a group of them came into possession of the location of the largest known diamond, although mining it may prove a little tricky – it's 50 light years away. The star, "romantically" styled BPM 37093, is a highly compact remnant of a white dwarf in Centaurus. Its incredibly dense sphere of carbon ash is now in a state of crystallization, forming a solid diamond 4,000 km (2,500 miles) across. Weighing 2 septillion tons, this is equivalent to 1×10^{34} carats! Discovered by a team of researchers at the Harvard-Smithsonian Center of Astrophysics, they decided to call it "Lucy" after the Beatles song "Lucy in the Sky with Diamonds."

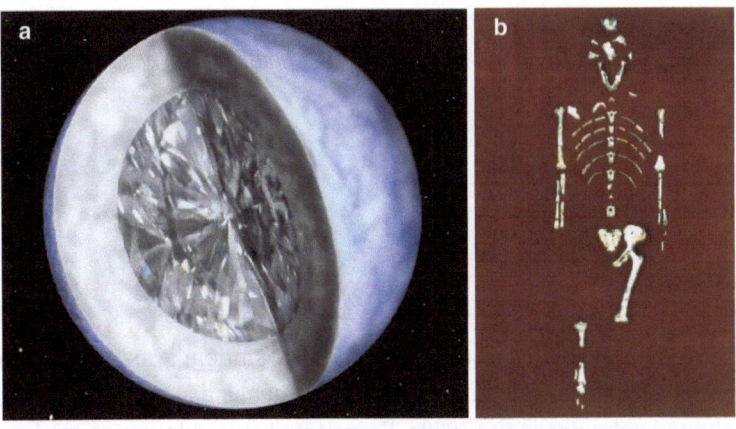

Fig 2.2 The two "Lucys" of science; a) BPM 37093 and b) the female Australopithecus afarensis [Images: Courtesy of T. Metcalfe (NCAR) & R. Bazinet (CfA) Harvard-Smithsonian Center for Astrophysics and Dr. Donald Johanson of the Institute of Human Origins at Arizona State University]

Oddly, this is the second of science's appropriation of this song's title. An early female hominid also bears this name, the tune having been on the air at the time of the skeleton's discovery. Thus is the eternal association between girls and diamonds perpetuated.

(VERY LITTLE REMAINS OF THE OBSERVATORY AND INSTRUMENTATION OF THE GREAT PRE-TELESCOPIC SIXTEENTH-CENTURY EUROPEAN ASTRONOMER TYCHO BRAHE. HOWEVER, THOSE WITH THEIR EYE OUT FOR HEAVEN-SENT OPPORTUNITIES STAY ALERT, SINCE UNKNOWN ANTIQUARIAN FINDS WOULD BE BEYOND PRICE.)

Dear Steve,

Clearing out my grandfather's attic last week, I came across an antique wooden frame roughly the shape of a quarter circle. It has tiny divisions along one edge that also has a barely discernible inscription saying 'Tychonis Brahe.' What do you think this is? I don't know anything about this astronomy lark, but it looks vaguely astronomical. Do you know who this Brahe guy is?

Homer
Syracuse, New York

Dear Homer,

This is without doubt a poor example of a student's astronomy project. Tychonis Brahe was probably a Greek lad involved in one of those student exchange schemes in the 1960s. Being totally worthless, you might simply throw it away. But as I might be able to find a use for it, and to help you with your clearing out, I'll take it off your hands. A suitable box is in the post. Please wrap carefully.

Tycho Brahe was the last great pre-telescopic astronomer who tragically died only months before the spread of the new-fangled optical thing. This sixteenth-century Danish nobleman was the first astronomer to build what might be termed these days as an observatory complex, although in style it was closer to a Disney-esque palace. Called Uraniborg, on the island of Hven off the coast of Denmark, it featured the very latest stellar sighting equipment.

Incorporating his own invention of the transverse vernier, Tycho's instruments were so accurate that the observations made with them were good enough for Johannes Kepler to calculate the orbit of Mars – and subsequently informed Kepler's three laws of planetary motion. Famously, only a few stones remain of his great observational edifice.

Tycho is cursed for all eternity to endure the vaguely smutty English mispronunciation of his surname.

(HAVING IN THE PAST BEEN INVOLVED IN OPENING UP MY LOCAL SOCIETY'S OBSERVATORY TO EVENING VISITORS, I HAVE OFTEN BEEN ASTONISHED – AND PERHAPS A LITTLE DISILLUSIONED – BY THE POSITIVE TURNOUT ON PUBLIC OBSERVING EVENINGS DURING RAGING THUNDERSTORMS. UNFORTUNATELY, THE ASSEMBLAGES ARE NOT BASING THEIR ATTENDANCE ON METEOROLOGICAL OPTIMISM, BUT ON AN UNREASONABLE EXPECTATION OF A TELESCOPE'S ABILITIES.)

Dear Steve,

I have recently been to a local observatory (over several nights) to look at Halley's Comet through a big telescope. I was unable to do this, however, because on each occasion it was cloudy and (though big) they said that the telescope could not see through cloud. Are they telling the truth? If so why is this?

Pete
Jackson, Mississippi

Dear Pete,

Visible light has a great deal of trouble getting through Earth's atmosphere as it is, without the added difficulty of trying to pass through banks of condensed water vapor and then (via telescope) penetrate the visual cortex of your troubled brain. (The latter constitutes the major obstacle.) Might one ask if you have X-ray eyes that see through walls into adjoining rooms?

Yes, this did happen to me. During Comet Halley's last apparition I was called to attend my local society's observatory where, believing the torrential downpour to be no obstacle, a family had arrived to take a look at the comet. While interestingly reflecting the medieval belief that comets were terrestrial vapors, it is a sad indictment of education that such conceptual errors can still exist.

It was sixteenth-century astronomer Tycho Brahe who, by parallax observations of the comet of 1577, showed that comets were not Earthly miasmas transporting humankind's sins towards heaven but real objects traveling through deep space, or at least further away than the Moon.

(SOME STARS ARE MORE MYSTERIOUS THAN OTHERS.)

Dear Steve,

I have always wondered what the Star of Bethlehem was. Do you know?

Cary
Lake City, Florida

Strangely enough, I believed that I saw it myself only this Christmas. I saw it hovering in the sky, not far from the giant constellation Orion. As I stood for a while and watched, Orion of course moved majestically westward. But this single brilliant star stayed exactly where it was, hovering beneath the celestial vault, completely immobile.

Thinking automatically of a second messianic birth, I decided to investigate. Walking towards it, I found it getting apparently higher in the sky, becoming brighter all the time. The closer and closer I got, the more its elevation and intensity increased, until at last it was so high that I had to crick my neck almost vertically to see its blurry image. It was then that I visciously struck my chin on the metal column that supported the streetlight. Believe me, for quite a while I saw many stars, all circling above my head. This had absolutely nothing to do with my prior indulgence at the local bar.

The real nature of the Star of Bethlehem has exercised the minds of scientists and theologians alike. Bright planetary conjunctions, comets, and stellar explosions have all been drafted in as explanations. Unfortunately some rather hazy dating is involved. This initially arises from a sixth-century calendrical miscalculation by the diminutive historian Dionysius Exiguus (Dennis the Small!). He calculated the year of Christ's birth by counting back through the reigns of the Roman rulers. Unfortunately, although he did extremely well, he was unaware that Augustus Caesar had also ruled as Octavian for 4 years. Compounded by the Romans' non-use of zero, this means that his calculation was off by 5 years, an error we still live with today. Christ's birth is now "guestimated" as occurring at around 4 B.C., which means he would have been about 5 when he was "born," according to Christian dating!

Fig 2.3 Star of Bethlehem [Image: Courtesy of www.from.old.books.org – from "Religion in the Home" by Charlotte Yonge]

While other dating difficulties arise from the problem of correlating ambiguous biblical texts with known historical events, the star's singular mention in only Matthew's gospel is also a concern. It should be borne in mind, too, that traditionally a messianic birth was required to have such an apparition associated with it for the occasion (and the applicant!) to "qualify."

Apparitional candidates include Comet Halley (12 B.C.), a triple conjunction (three occurrences) of Jupiter and Saturn (7 B.C.), conjunctions of Mars, Saturn, and Jupiter (between 6 and 4 B.C.), a lunar occultation of Jupiter (6 B.C.), a nova (5 B.C.), a comet (4 B.C.), and a conjunction of Jupiter and Venus (2 B.C.). Another suggestion is that it might have been an eruptive variable star, proposed by British astronomer Mark Kidger.

(THE SIMPLEST QUESTIONS ALWAYS TURN OUT TO BE THE BIGGEST.)

By email
From: Arnold Templer[templar@globular.com]
To: Dear Steve[doctorsteve@help.com]
Subject: The Universe

Dear Steve,

We are told that the universe we see around us is only a fraction of its true size – and that the universe is everything there is. But doesn't having 'size' imply there is stuff around it which is even bigger?

From: doctorsteve@help.com
Sent: July 30, 1991 09:45
To: Arnold Templer[templar@globular.com]
Subject: The Universe

Hi Arnold!

It's simple. So far as we know, before the universe there was nothing. After the universe has ceased to be there will be nothing. Therefore, by definition, 'nothing' is bigger than the universe.

As Douglas Adams put it, "Space is big, really big." As a primitive biped equipped with evolved cognitive abilities honed only for running around on the surface of a tiny spherical speck of stellar dust, all other appreciations seem superfluous. It is therefore an incredible achievement of intellect to even attempt comprehension of the limits of the universe – and even beyond.

Fig 2.4 There's more to the universe than meets the eye, as Empedocles discovers here in this medieval print [Image: Unknown medieval source]

Dear Steve,

I am six years old an veri Intrested in astromoly an I wood lik a bigg roket but mi momy says I hav got to sayv up I hav got nine centz can I hav wun plese.

Anne
Marion, Ohio

Dear Anne,

I am afraid that requests of your kind are *very* numerous. Because of this, Saturn V rockets, etc., are in very short supply. However I suggest that you send your nine cents to NASA at Kennedy Space Center. I think they are currently quite short of cash and may still have a few Apollo capsules left from their last sale. Good luck.

NASA's dwindling fortunes may indeed make it grateful for the 9 cents. Year to year budget cuts have steadily trimmed its aspirations in both manned and unmanned exploration. The International Space Station, despite being a major manned flagship, is only a shadow of the initial design concepts. Unmanned probes, reputedly "faster, better, cheaper" are in fact cheaper, cheaper, cheaper. (Although there have been spectacular successes.)

It is a sad reflection on our civilization that although scientists supply the aspirations it is politicians who supply the finance. Since politicians react to the pressure of their voters, it is up to all of us to ensure the money keeps flowing. It is sobering to consider that at the height of the "space race," NASA's 1966 annual budget was $27 billion; yet by 1987 it had fallen to less than $10 billion. The pressure of shuttle development and the International Space Station had squeezed this up a little, but it still hovers at about $13 billion (less than half its 1960s heyday!).

(IT IS A WELL-DOCUMENTED PHENOMENON THAT PURCHASE OF A TELESCOPE INITIATES A PROLONGED SPELL OF UNEXPECTED LOW CEILINGS, TEN-TENTHS UNREMITTING CLOUD COVER – IN EUROPE AT LEAST.)

Dear Steve,

I have spent lots and lots of money on acquiring a telescope and auxiliary equipment. I am ready and waiting to use it for the first time. But ever since I bought it three weeks ago there has not been a single clear night sky. Now every night I gaze hopefully up to the heavens, only to see a damn blanket of total cloud cover once again. I am fed up to the teeth with this happening. It's always cloud cloud cloud – and more cloud. Why the hell is it never clear? I might as well have saved my money or bought a train set for all the good my telescope is doing me. I know the sky must clear eventually, but how oh how can I improve my seeing conditions?

John, Swansea, Wales

Dear John,

Emigrate!

Cloud is possibly the best reason to get born on another planet, preferably one without damp aerosols. It is a celestial obscuration with a great sense of timing, occurring with uncanny accuracy during eclipse or lunar occultation observations.

Cloud, of course, is not the astronomer's only adversary. Differentiated air pockets in the atmosphere can make telescopic images wobble, boil, and blur. Faint but omnipresent high mist can dim objects below normal visibility with reflected light pollution, blotting out the fainter stars. Indeed, perfect observing nights allowing a telescope to perform to its theoretical limits may number only 5 or 6 per year. These nights, however, will hardly ever coincide with those occasions when astronomers are actually out observing.

Believe it or not (and ask any astronomer to confirm this) the problem referred to is a real phenomenon. Acquiring a new piece of astronomical equipment immediately initiates a protracted bout of cloudy weather. It's spooky but true. If you get a piece of astronomical equipment on approval, make absolutely sure that sufficient time is allowed!

(I CHALLENGE ANY AMATEUR OBSERVER TO DENY THAT EVEN AT THE FARTHEST REACHES OF THEIR MIND, THERE IS NOT THE FAINTEST HOPE

THAT THEY WILL MAKE A NEWSWORTHY OR EVEN GROUND-BREAKING DISCOVERY FROM THEIR BACKYARD.)

Dear Steve,

I have been observing for 25 years, out at all hours of the night, as zealous now as when I was just a lad. But at no time during this period have I discovered a new comet, star, or heavenly object of any kind. How can this be? Is there such a thing as 'luck' in this scientific pastime?

Ritchie
St. Louis, Missouri

Dear Ritchie,

You mean you don't know? Only by offering a blood sacrifice can failure be allayed. Unfortunately (unlike the Romans), the rather squeamish mores of today discourage the ritualized dispensing of living, breathing organisms (more's the pity!). However, I have it on good authority that the ceremonial cleavage of a particularly thick and juicy raw steak from the supermarket will do just as nicely. But before even thinking of building the observatory in your backyard, an altar for this purpose must be considered first. A word of caution; ensure that the sacrificial plinth includes an adequate fluid run-off into a suitable container. Congealed blood can be a trifle tacky underfoot and a pungent magnet for local Dracula wannabees.

Most discoveries in astronomy are down to hard work. The happenchance of a new comet or supernova may appear to be opportune, but behind the apparent serendipity, vast periods of dedicated time have been spent. George Alcock, legendary British discoverer of five comets and five novae, spent thousands of binocular hours scanning the sky armed only with a superhuman knowledge of the sky – which included memorizing 30,000 stars.

Of course, in the old days it was easier. When Galileo turned his telescopes to the sky, everywhere he looked presented him with new discoveries. The Moon had been a familiar, even hackneyed, object to humans since before the dawn of civilization. Yet in a trice it lost its silvery smooth perfection to be revealed as having a rough-hewn pockmarked scarred terrain. The stars themselves were joined by unsuspected lesser companions that far outnumbered those visible to the naked eye alone.

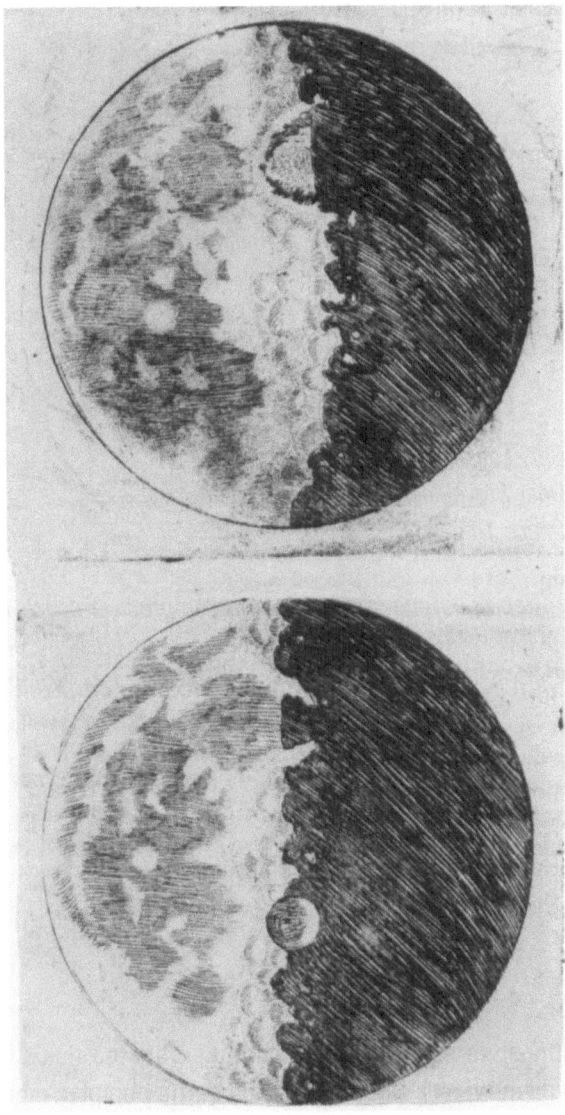

Fig 2.5 Galileo's rather poor lunar drawings [Image: courtesy of Siderius Nuncius 1610]

The planetary spark of light that was Venus revealed itself as a sphere displaying phases like the Moon. More significantly, Jupiter became the governor of its own domain of encircling planets. These last two discoveries alone were responsible for the downfall of an Earth-centered cosmology that had held sway for almost two millennia. All this, with a telescope whose aperture was less than that of the smallest and cheapest available today.

(IF SUCH INCENTIVE WERE EVER NEEDED. . .)

By email
From: Gary Louse[garyell@globular.com]
To: Dear Steve[doctorsteve@help.com]
Subject: Space Travel

My friends often upset me by criticizing my interest in space travel (which I talk about a lot), and I must admit to being hard pushed sometimes for a response. They say there is no point to space travel. What should I tell them?

From: doctorsteve@help.com
Sent: November 3, 1999 18:01
To: Gary Louse[garyell@globular.com]
Subject: Space Travel

Hi Gary!

First, the atmosphere of Venus contains carbonated fluids (in particular, CO_2 dissolved in water). Secondly, study of the complex molecules found in interstellar space reveals the presence of ethyl alcohol mixed with amino acids. For me, there can be no greater incentive than the inexhaustible supply of whiskey and soda!

Fig **2.6** The incentive for space exploration [Image by the author]

Justifying the cost of space travel can be hard at times. One can only say that as a hunter gatherer, exploration of our territory is in our genes. Having almost exhausted Earth's environment, what's left to do is out there.

The discovery of complex molecules in the interstellar medium began in the late 1930s. These now include formaldehyde, alcohol, ascetic acid, propanal, benzene, and nitrous oxide – although someone may be having a laugh with that last one. Methyl alcohol (methanol) was discovered in the B2 molecular cloud in Sagittarius in 1970. Regrettably, despite being an alarming intoxicant, it is fairly harmful. Fortunately, the situation was rescued in 1975 by the discovery of eythl alcohol (ethanol) in the same place. This is the stuff we put in our normal recreational beverages. Sadly, the B2 cloud is more than 25,000 light years away. The feverish hunt for a more accessible source continues!

The universe is veritably strewn with the amino-acid building blocks of life. Is our version of perambulating molecules unique? I don't think so.

(THE PRONUNCIATION OF A WELL-KNOWN COMET AND ITS HERO HAS BEEN, AND CONTINUES TO BE, A HOTLY DEBATED SUBJECT.)

Dear Steve,

I have trouble remembering how to pronounce Edmond Halley's name. Is it Hall-aye How-ley, Hal-ey, Harley, or Horley?

Sally
Glasgow, Scotland

Dear Sally,

I have the same problem myself, you know. I find it easier to refer to him as Grand Old Edmond, or better still, simply as Big Ed.

Why oh why did this eighteenth-century English polymath inconsiderately have his name attached to a comet without telling anyone how his surname is pronounced? The conventional wisdom nowadays (although somewhat contrived in my humble opinion) is to pronounce the man's name as HAW-ley and the comet as HAL-ee.

It is often forgotten that the tentatively predicted revisit of this comet (in the nick of time) at Christmas of December 1758 confirmed Newton's new theory of gravitation. Using its principles, Halley conjectured that the bright comets of 1531, 1607, and 1682 were in fact the same comet. He went on to predict its return in 1758. Although Halley died in 1742 and therefore was presumably unable to enjoy its return, he is nonetheless honored with its naming.

It is an interesting irony that although comets are described as the nearest that anything can be to nothing and still be something, they hold clues to the creation of the Solar System itself.

(THE SKY IS A BIG PLACE. FINDING PARTICULAR OBJECTS OF INTEREST CAN BE HARD. YET A METHODOLOGY CALLED STAR-HOPPING CAN BE USEFULLY EMPLOYED. SOMETIMES.)

Dear Steve,

I would like to observe a rather little known nebula known as the scarlet hoopla (XC49Ce .5-502-z9/TR). Could you please tell me where to find it?

Randolf
Phoenix, Arizona

Dear Randolf,

Delighted! First, start from the well-known square of Pegasus, and its top right star Scheate. Follow a line up past Andromeda until you reach Alpha Cassiopeia, turn sharp left, and halt at the Perseid star of M(b). Wait three minutes, then head east and up towards Alpha Camelopardis in a sort of wiggly line until you find one of the twin stars of Gemini (Beta Geminorum), Pollux to you. Bear a severe right angle until you come across a bluish twinkly star with just a faint hint of purple and trace the indistinct line of stars nearby halfway down. Hold your hand at arm's length, placing your thumb on the spot, and scribe a short arc with your little finger. Bisect this angle and extend the line backwards for about 15 degrees. Place your (now) broken arm in a sling. Join this last position with the star you passed four minutes ago (taking care not to alter your line of sight), and imagine a spot two-thirds from each end. Lean your head left and anticlockwise, close the right eye, and then, if you know where you are, please tell me, because I'm damned if I know.

Being able to find astronomical objects is a desire that immediately follows knowing they are there. Leaping from easily identifiable objects (such as the bright stars of Ursa Major) to finding more difficult fainter targets is a craft learned by every novice astronomer. Called star-hopping, it forms an essential part of the process that turns the unfamiliar mass of the night sky into a playground of warm familiarity. Almost all beginner books describe how it is done. Generally, it is quite safe.

Of course, the sky might easily have looked entirely different. Although Johann Bayer issued one of the best and most beautiful star atlases in 1603 (it was he who kicked off the Greek letter stellar designations), another astronomer had a different idea shortly afterwards.

Julius Schiller was a devout Christian who, during the 1620s, became convinced that the sky should be depaganized. To this end, he devised a sky populated by biblical figures. Draco became Massacre of the Innocents, Hercules was changed to the Three Magi, and Pegasus was transmogrified into the Archangel Gabriel. The twelve signs of the zodiac were replaced, of course, by the twelve apostles. After many years of exhausting work he published his "Coelum Stellatum Christianum" (the Christian Stellar Sky), in 1627.

Apart from the revolutionary (perhaps revelationary) idea and fantastic artwork, he also took the trouble to represent the constellations in his atlas reversed, as they would appear on a globe. To his obvious consternation the idea bombed. He died a disillusioned man.

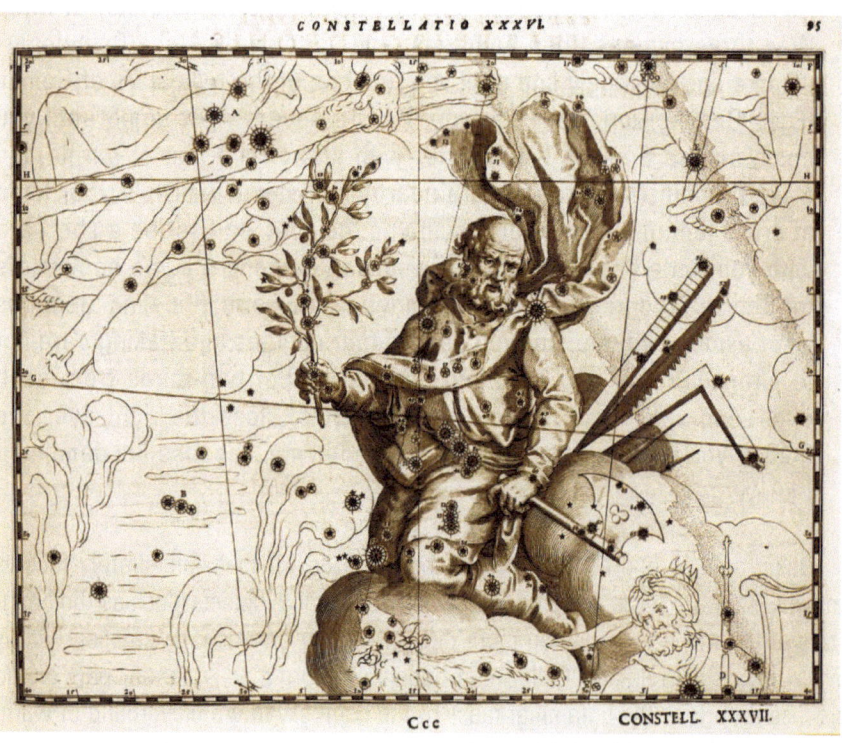

Fig 2.7 Schiller's Orion, Christianized into St. Joseph! [Image in *Coelum Stellatum Christianum* by Julius Schiller, 1627]

Just a few years ago there was another fright that the sky was about to change. It was announced to a shocked astronomical community that the European Community had plans to metricate the sky. The zodiac, we were told, was being changed from 12 to 10 constellations! The impish British astronomer Sir Patrick Moore was the source of the rumor, announced on April 1st (April Fool's Day).

(I ADMIT, I HAD TROUBLE WITH THIS, TOO.)

Dear Steve,

You must get lots of problem letters from novices who find some of the concepts in astronomy difficult to grasp. I've seen you help those with very thorny incomprehensions indeed get their brains to understand something. Mine is simple. But let me start by explaining where I'm coming from. Listen. When you make something bigger, it gains weight. Golly, I ought to know. I used to be 130 pounds; but now that I'm bigger and heavier I weigh nearer 200. OK, I take the point that I may need to diet, but the point I'm making here is that the numerical value of my weight has gone up. I got a maple tree in my back yard. Last year it was 11 feet tall – and this year at 12 feet it's bigger and taller – the number expressing its height has increased 'cos it's bigger. If I want to replace a 60 watt light bulb with a brighter one, I get a 100 watt. In other words the bigger the wattage the brighter the bulb. So, for gosh sakes, why is it that for goddam stars, it's the other way around; the bigger the number, the fainter they are! I just don't get this at all. I mean, that really bright beacon of the northern sky, Vega (in Lyra), only has a visual magnitude of about zero. Yet, the spiral galaxy M109, which is so faint you need a big telescope to see the damn thing, is listed as magnitude 10! What the hell is going on? If it's so faint, it should have a much teenier number than Vega. I mean, if this almost invisibly faint galaxy gets a 10 assigned, the bright star Vega deserves 500, surely? Obvious! Or rather, it's not. Could you explain this, Steve, before my mind just explodes? And it'll help me be a mite less argumentative, too.

Sam
Rockland, Maine

Dear Sam,

Take it easy! It's OK. It has to be said you are not alone in your consternation, so let me calm your troubled waters. This numerical bias you accept as the norm (of more equaling a higher number) is not as prevalent and dominating as you appear to think it is. Maybe, if I can redress this balance, you might regain a sense of balance for your ..er..um..unbalanced mind.

Let's look at a journey, any journey, to a destination. The more miles you travel towards it, the smaller becomes your trip. A larger value traveled results in a smaller number of journey miles to go. Get my drift? How about camera exposures? If you go from a 125th to a 250th, the number is getting bigger – but the exposure is getting shorter. So, shorter equals higher number! Counter-intuitive, or what? Tackling mathematics head on, let's discuss an ellipse. As its minor axis gets smaller compared to its major, the value of its eccentricity gets larger. So, the shorter it gets, the higher this value. Am I helping much? Getting closer to home, what about telescope magnification? If we double a magnification from x100 to x200, the area of sky you are looking at consequently gets smaller. So the higher the magnification, the smaller the angular area! Feel better? I hope I have explained to you that in many spheres of calculation, as something gets bigger one of its values may diminish. If you have not understood this by now, alas, there are just some things you may simply have to accept. Live with it!

I meet many amateur astronomers who find the stellar magnitude system counterintuitive. They can blame the whole damn thing on a Greek thinker, Hipparchus, who about 150 B.C. was indeed thinking about how a star's brightness could best be described and graded. Unfortunately for confused modern-day observers, his idea was accepted and "set in stone" by his fan Ptolemy 300 years later. Ptolemy's book, which later scholars would dub "the Great book" (in Arabic, al-kitabu-l-mijisti – or Almagest, for short) laid the foundation of the magnitude system astronomers use today.

Understanding its concept is actually easier than one might suppose. The clue is in the name. Magnitude! Magnitude literally means greatness – how great something is. Try replacing this word with a similar one, say "grand." The very brightest star could be viewed as being of the highest, or first, grandness – or put another way, a magnitude rating of 1. Stars that are less bright do not attain this marvelous pinnacle of luminosity and are only of secondary importance, secondary grandness; i.e., a (secondary) magnitude 2. Can you see where we are going with this? Hipparchus (and Ptolemy after him) decided to split all the stars into six groups of steadily decreasing magnitude (or lessening grandeur), with the highest level of 1 being the brightest and those at the lowest level (several steps down from the rostrum, so to speak) as of only 6th brightness. Astonishingly, scientists later discovered that unbeknownst to these ancient sages, the five divisions between the brightest and faintest stars visible to the eye correlated very closely to a logarithmic scale. This lucky happenstance was thenceforth formalized, with each whole number magnitude value being divided from its neighbor by the incremental brightness of 2.5. The original Greek subjective grading probably arose because the eye indeed reacts logarithmically to light. The stellar magnitude system was built-in to our optical physiology!

As we got better and more objective at measuring stellar magnitudes the archaic values handed down through history had to be refined. Unfortunately, the very brightest stars came out with reappraised luminosities "higher" than 1. This is why we now have stars that not only possess magnitude values of between one and zero (don't panic yet, it gets worse) but penetrate through zero into minus figures; Sirius, for example, shimmers at a marvelous −1.46. The Moon and Sun, using the same luminosity scale, burrow deeper still into the negative territory of the magnitude system. Perhaps, for the sake of sanity, we won't go there!

No, maybe we should. After all, the sanity of my readers is not my concern. The apparent magnitude of the Moon is −12.7, whereas the source of its light, the Sun, boasts a brilliance of −26.7.

CHAPTER 3
Instrumental Hiccups

If only he knew what he was starting, that long ago (and still disputed) inventor of the optical aid now known as the telescope. It was simple in the beginning – as all things are. A lens clutched in each hand, held one behind the other, to gaze longingly through the portal of a lady's bedroom.

But people started mucking around with it. Extra lenses were added to improve the clarity, magnification, and field of view. Monstrous supports grew up of increasing complexity to simplify the chasing of sky-embedded wonders. Soon, parasitic devices were invented to squeeze every drop of observational blood from the telescope. Even the vulgar light itself was forcibly disembodied by spectroscopes to reveal its inner nature. Thus was born the telescope accessory, the astronomical version of hi-fidelity's woofer or tweeter.

The accumulation of these accessories can creep up upon the unwary astronomer over many years. Before long, the total mass of these extra accoutrements can outweigh the telescope itself by several tons.

As for the workhorse itself; here, at least, we have a case where size really does make a difference. Despite the fact that the full glory

S. Ringwood, *Astronomers Anonymous*, DOI 10.1007/978-1-4419-5817-4_3,
© Springer Science+Business Media, LLC 2010

of a night sky can best be enjoyed by the eyes alone, astronomers ironically spend much of their time acquiring increasingly bigger bits of equipment enabling them to magnify progressively smaller fragments of it.

Generally speaking, the bigger your instrument, the more detail you will see. In the smaller ranges of telescopes, particularly, even an extra ½-inch counts. In how many areas of human endeavor can that be said?

Thus is born a commercial arms race of telescopic devices that will claim to allow perceptions that defy belief and the laws of physics. This shamelessly feeds the competitive spirit that underlies the thin and deceitful veneer of every co-operative observer. Plough on then, through the techno-babble; discover the awful truth at the center of every astronomer's black heart.

(THE LAST FIFTEEN YEARS OR SO HAS SEEN A QUANTUM LEAP IN THE USE OF LEADING EDGE TECHNOLOGY IN TELESCOPIC EQUIPMENT. THIS IS MAINLY COMPRISED OF ADVANCES IN ELECTRONICS AND LENS MANU-FACTURE. SOME OF IT ACTUALLY CONSTITUTES AN IMPROVEMENT. BUT, AS THEY SAY, "THE BIGGER THEY ARE...")

Dear Steve,

I have just purchased a sophisticated 10" Schmidt-Cassegrain telescope. It incorporates a seek and find auto comet finder, an infrared theft alarm system, a plutonium decay power pack, a double-wedged fully boggled whim-wham, a 1,200 lb retro-thruster backup for the anti-gravity field stabilizer, fire-basted spittle-coated optics, virtual reality eyepieces with focused diatomic ni-cam ninja digital stereo pupiliary sockets, a phallic flimmer for the rogerbotter, an integral accessory tray capable of holding the New York Philharmonic Orchestra, a handheld laser-pulsed control unit with 15 LEDs (5 of which actually indicate something), a strap-on cuddly toy for cloudy evenings, a chrome-plated infinite-simally adjustable nut and bolt, and a pre-stressed cylindrical tube decked out in a fetching shade of diamante lilac. I can even see stars with it. Unfortunately, all is not what it should be.

The gearings of the Right Ascension drive unit (responsible for star-tracking) do not mesh properly, giving rise to a wobbly effect that I believe is called backlash. Can you help?

Marie
Dawson Creek, British Columbia

Run the R.A. drive while feeding the ingoing thread with a well-chewed wad of the oldest piece of pavement chewing gum you can find. Add chocolate to taste. Failing this, you could try thumping it.

Trouble with new telescopes is nothing novel. Consider the disaster that was the Great Melbourne Telescope.

Fig 3.1 The Great Melbourne Telescope [Image: Mt Stromlo Observatory, The Australian National University]

In the mid nineteenth century the southern hemisphere had not yet benefited from the presence of a world-class telescope. Driven by grandee UK astronomers Lord Rosse and Armagh Observatory's Romney Robinson, an advisory board grandly called "The Southern Telescope Committee" was formed in 1852.

The well known telescope maker Thomas Grubb placed a bid to make a 4-ft mirror. Unfortunately the Crimean War delayed the project by 9 years. There followed arguments about the mirror's material (speculum metal against the new revolutionary glass substrates) and the type of mount (alt-azimuth or equatorial). There was even discussion on using instead William Lassel's gargantuan 4-ft aperture telescope.

Thomas Grubb finally got the go ahead on making a speculum metal 48-inch mirror. As he was busy, the job of casting the blank of this behemoth went to his 21 year old son, Howard. While getting the metal up to heat for the casting there was a spillage. The resulting fire damaged the factory so badly that it had to be rebuilt. One can only imagine how son told dad of this little mishap! But by 1868 the mirror was completed, varnished (for protection against the sea air), and shipped to Melbourne, Australia, where the body of the telescope had meanwhile been in the throes of construction.

When the mirror was unpacked at Melbourne, it was found that the acid flux used as sealant for its container had dripped on to the mirror and burned pockmarks into it. They also found they could not remove the protective varnish. The observatory itself was faring no better. The telescope piers constructed to support the tube were found to be the wrong size, and the mounting had been engineered to the wrong latitude. Also, because of poorly designed mirror supports, stellar images were found to be distorted by the consequently deformed mirror. The observatory, designed to be completely open to the elements, allowed the wind to shake the telescope. In 1870 the astronomer in charge resigned.

At the time, G. W. Ritchie declared that "I consider the failure of the Melbourne instrument to have been one of the greatest calamities in the history of instrumental astronomy."

So the next time one of your new scopes misbehaves, think yourself lucky. It could be worse. Far worse.

But here for once is a tortured story with a happy ending! Phoenix-like, the telescope found new life when it was acquired for the Mt. Stromlo Observatory in 1955, where it was installed in a new dome. It did sterling world class service there for almost 50 years, enjoying technical upgrades particularly in the latter part of its life, when it was a major resource in the search for the "missing mass" of the universe. Its renewed good fortune finally ran out on January 18, 2003, when it was engulfed and irrevocably damaged by a firestorm. A romantic, if tragic, end for a phoenix.

(MANY ASTRONOMERS ARE OBSERVERS OF THE OLD SCHOOL, ESCHEW-
ING MODERN TECHNOLOGY FOR TRADITIONAL METHODS.)

Dear Steve,

*Do you think the increasing replacement of eyepieces by CCD video monitors on
big telescopes is a good thing? I only use eyepieces on my amateur scope and am
scared to death of moving up to these devices.*

Jacob
Dixon, Illinois

Dear Jacob,

**I have absolutely no time for the damn things! I came unstuck one night in
the Canaries. At the observatory atop La Palma on the Roche de la
Moustaches I mistakenly thrust my trusty Huygenian eyepiece into a
fiendishly disguised power point – with shocking results!**

The great thing about charge-coupled devices and similar detectors is that they
are capable of delivering a telescope's raw image to a computer capable of enhan-
cing the original visual image. High demand has now enabled their cost to fall to the
point where they are now more affordable.

Unbelievably the first CCD's were designed in the early 1970s for use as a mass
storage device in computing. The year 1974 saw their first appearance as imaging
components – although the initial format offered a chip of only 100 × 100 pixels. Early
versions were hardly better than celluloid film in sensitivity. These days, of course, CCDs
have far greater efficiency, although there is still a bit of catching up to do on the
resolution front. Professional astronomers have been quick to take advantage of the
technology, on the ground and in space. Consumer-driven domestic CCD development at
one point even outstripped the detectors used in observatories – enabling a brief
photographic superiority by amateur astronomers – at least on the camera front. The
professional guys have since redressed this imbalance with a vengeance and now really
kick-ass with their multi-megapixel detectors. I do know some amateurs who still use film.

The Huygenian eyepiece mentioned here serves to remind us that technical
innovation is not new. Constructed from two plano-convex pieces of glass, this
ancient eyepiece design was invented by Christian Huygens in the early 1660s to
supersede Kepler's bi-convex single element ocular, itself the first improvement on
the bi-concave versions first used in telescopes. The Huygenian thus represents

humankind's first foray into multi-element (compound) eyepieces. Together with his general optical improvements to the telescope, this eyepiece allowed Huygens not only to discover Saturn's satellite Titan but also helped resolve that planet's "appendages" into an encircling ring.

Adding progressively more corrective lenses to the design of eyepieces is a trend taken to excessive lengths these days! Various kinds of eyepieces now have 4 (Plossl, orthoscopic), 5–6 (Erfle), and 7 (Nagler) elements. Some exotic wide-angle eyepieces now incorporate up to 9 elements. These last are only possible through high-refractive glasses and multiband multicoating.

Most of these extra elements are pressed into the singular service of allowing incredibly wide apparent fields, some reaching 100 degrees. Yet the human eye's central area of concentration is only about 25 degrees. Discuss!

(ARGUMENTS FLY CONTINUOUSLY BETWEEN ASTRONOMERS CONCERNING THE VARIOUS MERITS OF DIFFERENT TELESCOPE DESIGNS. THERE APPEARS TO BE NO DEFINITIVE ANSWER, BUT ALL ASTRONOMERS HAVE THEIR OWN FAVORITES AND DISLIKES.)

Dear Steve,

I have become deeply worried recently about my feelings towards another type of telescope. All my life I have read about, built, and used Newtonian reflectors. But lately my thoughts have been turning to catadioptric systems such as the Schmidt-Cassegrains. However perverse this sounds, I find I cannot help it. I have tried not looking at the ads for them, avoided looking through them at star parties or even talking about them. Unfortunately, the harder I try to keep on the straight and narrow, the more intense my interest in them becomes. I find it hard to admit to myself that I might be becoming a Cat' lover, but this seems to be my destiny. I thought I might try professional help. Perhaps some aversion therapy will do the trick. What should I do?

Anon.
New York, NY

Dear Anon.,

I know from personal experience how painfully difficult this obsession can become. In fact I know several friends of mine whose secret leanings have been exposed by the Multi Element Optics Group. So far as treatment is concerned, I'm afraid it is doomed to failure. I do, of course, know people who have tried hanging heavy weights from their tripods, going to bed with their current instruments under the pillow, thinking Newtonian thoughts, when abhorrent desires arise. But these measures only serve to delay the inevitable – dare I say it – the actual purchase of one of these compound instruments. I can only sympathize. I am sending you the contact number of a help group in your area of those similarly afflicted.

The catadioptric telescope (an optical system employing both mirrors and lenses) is used extensively by telescope manufacturers because it can get an optically longer telescope into a smaller box.

Bernard Schmidt devised an early form of catadioptric telescope in 1930, but since the point of focus was situated deep within the tube it was only suitable for photographic use. The design breakthrough for visual use was devised by Dmitri Maksutov in 1944; but the Cat' family has since expanded considerably, notably including the modern-day Schmidt-Cassegrains. These telescopes are popular because they are compact for their focal length and (from a manufacturer's point of view) have optical designs that suit them to mass production.

There can be no doubt that the growing interest in astronomy is fueling a continual improvement in commercial telescope construction. With new types of

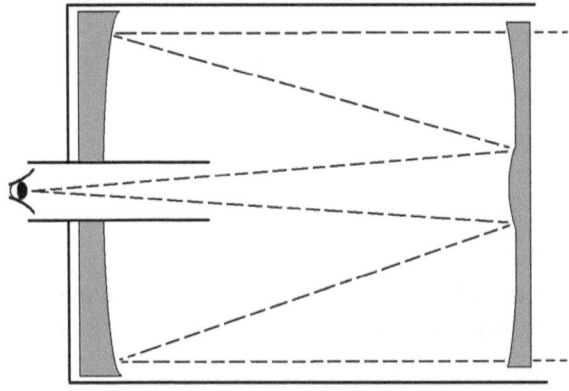

Fig 3.2 Dotted line shows the light path in a typical catadioptric design – the Schmidt-Cassegrain telescope [Image by author]

glass being devised all the time, optical designers are perpetually fed the raw material for design improvements. But basically it is horses for courses. Particular designs will always best match specific uses. Different telescopes play their part as the dumpster trucks (reflectors), racing cars (refractors), and Jeeps (Schmidt-Cassegrains) of observational astronomy.

(IT IS ONLY FITTING THAT IN A SCIENCE OF VISION, THE GREEN-EYED MONSTER IS ALWAYS PRESENT.)

Dear Steve,

A pal of mine has recently bought a telescope that is deliberately an inch greater in aperture than mine! He is now always bragging how much clearer his views are and takes every opportunity to scorn my observations and photographs whenever I show them at the local astro society. He has even taken the trouble to calculate the relative differences between our two instruments so that he can say (in my earshot) that his light grasp is x percent greater than mine, or that my contrast and resolution is a mere y percent compared to his. I used to be the most popular person at the local society, a position that he has now usurped. He now always has a crowd of people in attendance, lapping up his latest exploits. Urgh! Am I being unreasonable?

Ted
Seattle, Washington

Dear Ted,

It is a largely unknown fact that mono-anhydrogenous dimethyl ammoniumnitrate (MaDmAn) is an untraceable compound that, though harmless to humans, has the delightfully delayed effect of dissolving optics. I suggest that you acquire a bottle of this substance (address in the mail) and present this to your 'friend' as a gift of lens cleaner for his new telescope. Sit back and enjoy!

The aperture of a telescope (i.e., the size of the light-collecting element) governs its abilities in two ways. The first, its ability to show detail (or resolution), is controlled by the aperture's linear diameter. This is normally how a telescope is defined (i.e., an 8″ reflector). The second, light-gathering power, is determined by the total light-gathering area. Although a small increase in aperture affords little improvement in resolving power, it provides a disproportionately larger improvement in light grasp (since the area within a circle rises much faster numerically as you increase its diameter). This innocent property of Euclidean mathematics is entirely responsible for the acquisitive drive known in amateur astronomical circles as Aperture Fever. This is a manic craving for ownership of ever-bigger telescopes, an incessant and incurable condition manifesting itself as the iterative purchase of increasingly larger instruments – and abject poverty.

Apparently diminutive spatial differences between telescopes can translate into a large divergence in ability. However (with a big bold capital "H"), it is the observer behind the eyepiece that can make the greatest difference. A trained eye will see much more with a small telescope than a novice with a larger instrument. This "eye training" can take some time – a point that should be borne in mind by every new telescope owner.

The problem featured in this "Dear Steve" can be observed running rampant at every star party, where astronomers show off their latest instrument to a gathering of other astronomers doing the same!

Those who want to play this game may find the following formulae useful. Resolution (in seconds or arc) is determined by dividing 4.54 by aperture inches, or 116 divided by aperture millimeters. This determines the closest distance in arc seconds (Dawes' limit) by which stars need to be separated to show as single stars through a telescope. Do not confuse this with Rayleigh's limit, which is the degree of separation sufficient to betray stars as doubles. Limiting magnitude (the faintest stars visible) changes a little with magnification but is generally derived from 7.5 + 5 log D, where D is the telescope's aperture in centimeters. Enjoy.

(OWNERSHIP OF A TELESCOPE CAN DEVELOP INTO A DEEP AND CARING RELATIONSHIP, THOUGH PERHAPS A SOMEWHAT ONE-SIDED ONE. IT IS NO SURPRISE THAT STRESS CAN RESULT.)

Dear Steve,

I hope that you can offer advice on a decision I have to make soon. My lovely telescope has lost its ability to point. I have been told that an operation is necessary, otherwise the condition will deteriorate to the point where the 'scope will have to be put down. It is a complicated procedure involving 'the bearings' (apparently, a serious operation). I have been told that the chances of it pulling through are 50/50. Do I wait and risk the condition worsening to the point of no return, or do I put it through the operation knowing that it may hasten the end? I have had the telescope for a long time, and it grieves me terribly to see it suffering in this way. I have even considered telenasia. I feel sure it knows there is something badly wrong, and of late I cannot bear to look it squarely in the lens (although I try to appear cheerful for its benefit).

Henry
Cork, Ireland

Dear Henry,

I can tell from your rather damp letter that you are really upset. Indeed, one must do a great deal of soul searching on these occasions. It really depends on whether, for sentimental reasons, you wish to allow your 'scope to continue its decline in a dignified way or risk dismantlement to give it a new lease of life. I generally take the view that remedial treatment, though beneficial in the short term, can only delay the inevitable. Perhaps you should lay your faithful friend aside for a well-earned retirement in a dark cobwebbed corner somewhere. Then go for it! Buy yourself a super spanking new upgraded instrument to take you further in your observation of the universe. (This is not as insensitive as it sounds. Now stop crying!)

What to do with your old equipment when you "move on" to bigger or better stuff is a question that is on occasion surprisingly tortuous. I know many astronomers who simply pass on used telescopes to others – particularly if doing so provides funds towards the new telescope. Yet be warned. I gave away my very first telescope without realizing the emotional scar its loss would inflict. (Don't despair. The therapy helps.)

The dangers inherent in sentimentally retaining equipment too long cannot be overstated. The 300-ft radio telescope at Green Bank, Virginia, had put in many long years of faithful service since its construction in the early 1960s. One of its most famous tasks was that of pointing towards sun-like stars and listening for extraterrestrials. Unknown to the users of the telescope, almost three decades of continuous maneuvering had taken its toll. During an observing run on November 15, 1988, the whole edifice simply collapsed into a heap of scrap metal! Miraculously, no one was injured.

Fig 3.3 Green Bank – before . . .

Fig 3.4 . . . and after [Images courtesy of NRAO/AUI/NSF]

A subsequent investigation of the cause found that the original formulae used to calculate the engineering stresses had since been superseded by new ones that would have predicted the disaster. But there was a silver lining; the collapse instigated the construction on the same site of the Robert C. Byrd Green Bank Telescope, a unique 485-ft-tall world class radio telescope.

Outmoded design may also have been a factor in the collapse of the 30-year-old Hat Creek Radio Observatory in California in February 1993. The dish fell during a storm following two 100-mph bursts of wind.

Fig 3.5 Hat Creek Radio Observatory, the 85-ft dish lays sprawled at the foot of its tower, the base of which can be seen on the left. [Image courtesy of Richard Plambeck, University of California at Berkeley]

(GUIDING LIGHT...)

Dear Steve,

I am having great difficulty in aligning the mirrors of my Newtonian reflector. Is there any literature that can assist me in this task?

Daniel
Thunder Bay, Ontario, Canada

Dear Daniel,

Try reading the excellent book on this subject, *Bending Light to Your Will* by Professor Colin Maishon, published by. Turnbolt Slowly Productions.

Collimation of telescope optics becomes more critical the lower the focal ratio of the instrument. Alignment of the primary lens of an f15 refractor is therefore quite forgiving, yet is desperately critical for an f4 reflector. No telescope will perform to its theoretically maximum abilities unless it is correctly collimated.

Many devices are available to make this infinitesimal adjustment visible; most commonly is the one known as the Cheshire eyepiece, which can be used to good effect on any telescope design. (See Social Torments, for more on collimation) tips.

(ONE CHARACTERISTIC PARTICULARLY SCREAMS FROM THE PAGES OF ANY AGONY COLUMN. PROBLEMS THAT ARE SMALL IN ACTUALITY TEND TO BE WRIT LARGE WITHIN THE TORTURED MIND OF THOSE AFFLICTED.)

Will the young man who wrote to me about the problem he is having with his rack mount please send me a self addressed envelope so that I can send him details of the various plants that can remedy his difficulty.

The rack mount – the adjustable tube that controls the position of an eyepiece – is without doubt one of the most important components on a telescope. Its delicate motion has a direct influence on the sharpness of a telescope's focus. If prior to purchase you can only test a telescope in a store, make this assessment a primary one. If its motion is at all floppy, juddery, or variable in tension, know that these deficiencies will be magnified a thousand-fold in use.

The earliest telescopes were of course simply push-pull tubes – which must have been fun on those rickety pillar supports. But by the end of the seventeenth century rack and pinion focusers were almost universal.

The problem with rack and pinion gears, in which a toothed wheel drives a linear track, is that there is commonly an inevitable (albeit brief) loss of contact with the rack at some point during rotation of the wheel, although precise engineering can minimize this "backlash". This remained the case until the mid-1960s, when John Wall of the UK's Crayford Astronomical Society devised a solution.

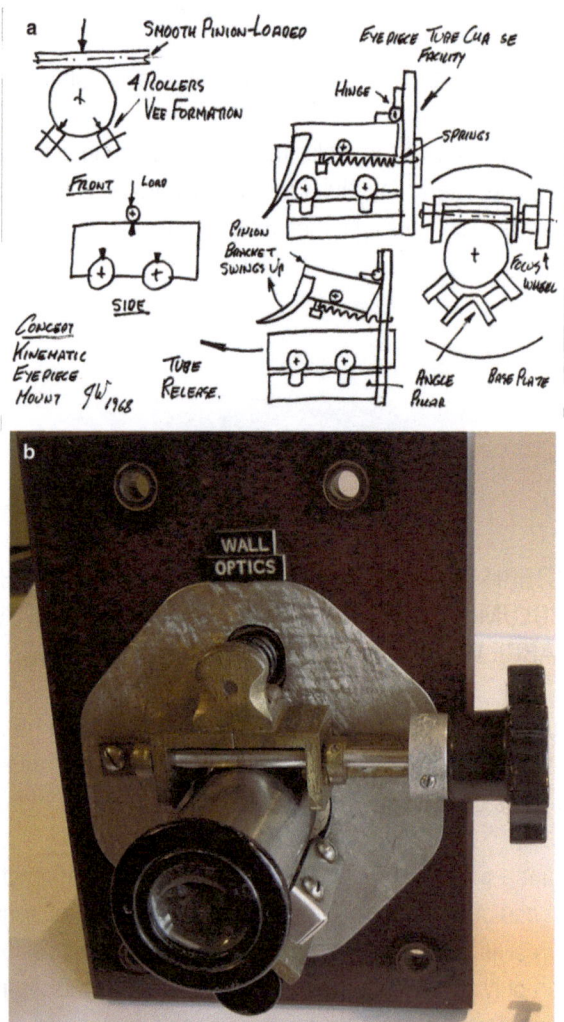

Fig 3.6 (a/b) The Crayford; a revolution in focusing a) concept sketch, b) finished product. [Image courtesy of John Hall]

Wall was wrestling with the problem of needing a focuser that could deal with the critical focus adjustment he required on a 34 cm f4 comet seeker he was constructing. In a feat of brilliant innovation, he devised a friction-driven eyepiece tube supported essentially by a roller bearings system. The design, although initially slow to spread, became popular because it was incredibly smooth, accurate, and comparatively easy for amateurs to construct. Retail retro-fit versions began appearing in the early 1970s, and the Crayford eyepiece mount can now be found on many, if not most, commercial telescopes today. Hope he took out a patent! Wall named it the Crayford eyepiece mount in honor of the town of his birth and the astronomical society he co-founded there. Just as well, perhaps. Had he named it after himself, it would have been called the Wall Mount – indicating an entirely different mechanism!

(WE SHARE THE WORLD WITH MANY WONDROUS CREATURES.)

Dear Steve,

I have a problem that few astronomers I'm sure have to deal with. Every time I use the 10-inch Newtonian reflector in my observatory, earthworms slither up the instrument and slide around inside the eyepiece. A most disagreeable experience. And the slime they leave. Ugh! How can I stop them?

Rachel
Hopedale, Newfoundland, Canada

Dear Rachael,

Good heavens, girl! Don't you realize this is evidence of intelligent pursuit? Burrowing beneath your observatory over many years, generations of them have been so closely exposed to the wonders of the cosmos that they have now evolved to a stage of scientific endeavor. They now wish to use your equipment to view the universe. Yet, I cannot help but wonder if you have your telescope mount rather too close to the ground.

I sympathize with your objection, as slime is very bad for optics. I suggest you set up an observing session during a clear day and point the 'scope at the Sun without a filter fitted. The precocious creatures will be incinerated as they try to make an observation, discouraging further occurrences. But the smell. . .the smell. . .

This letter was inspired by a tale concerning the late, great science fiction author Arthur C. Clarke. He had been left observing Mars through a big professional telescope when, late in the night, an attendant heard him crying out "The Martians are coming! The Martians are coming!" The attendant, rushing to the scene, found a crazed Arthur gesticulating to the eyepiece with great alarm. Upon looking through the eyepiece himself, the other observer saw that a fibrous cross-wire in the optical path had been incinerated by its hot illumination. The resultant embers were floating off down the telescope tube. Grazed by Martian light, they looked like an approaching invasion fleet taking off from the planet. Both men fell about with laughter.

Unfortunately (despite a brief correspondence), I never met Arthur C. Clarke, who sadly died in March 2008. But in my one-time meeting with his brother, Fred, I found out that he, too, exhibited a well developed sense of humor.

(AMATEUR ASTRONOMERS ARE NOTORIOUS FOR THEIR INGENUITY IN SOLVING ENGINEERING PROBLEMS.)

Dear Steve,

I like to attend star parties and observation evenings in my locality, but cannot do as I would wish – which is to take my 12" reflector with me. The car's too small to carry it. Do you have a solution?

Jock
New Glasgow, Nova Scotia, Canada

Dear Jock,

You are looking at the problem from the wrong viewpoint. Which is more important, the car or the telescope? Obviously it is the latter. Since you ask me for a solution, can I suggest a little re-designing of the reflector tube? Lay the tube on the ground. At each 'corner' fit a small axle and wheel of sufficient diameter to raise the instrument a few inches off the ground. Midway along the top of the prone tube a comfy cushion can be strapped. Now, sitting astride your new 'vehicle,' you can propel yourself to any observing event by pushing with your feet; or even be towed by a (hopefully) considerate friend.

Astronomy is certainly one of the few sciences where you can, if desired, make your own equipment – telescope, mounting, accessories, the lot! There are almost as many books on making your own observing equipment as there are on astronomy itself. It is often a far cheaper option, too.

I know colleagues who cannot walk down a road and pass by discarded utilities tubes without wondering where they could find a lens to fit them. I have known projector lenses, spectacle blanks, and even components from photocopiers to be pressed into such service.

Of course, it is one thing to turn scrap into a telescope, but quite another to turn a telescope into scrap. Yet this is what happened in 1836, at the conclusion of the debacle in the UK between Sir James South (President of the Royal Astronomical Society, 1829–1831) and internationally renowned telescope maker Edward Troughton.

South began to build an observatory in Kensington, West London (no light pollution in those days!). In 1829 he acquired the largest refractor lens then available (12″) and set about the construction of the dome and mounting for this huge telescope.

He commissioned Troughton to construct an English mounting, a design Troughton warned was unsuitable for a telescope so heavy. Troughton's fears proved true. The mounting's polar axis vibrated each time the instrument was turned – apparently twitching 10 or more times with an amplitude of half a second each. This was clearly unsuitable for use. Thus began an increasingly acrimonious and public dispute between South and Troughton.

Assisted by other astronomers and engineers, a series of design modifications were gradually introduced. The telescope was by stages declared fit for use. But not by South. He refused to pay for the work being done – despite the fact that the cost to Troughton had long since exceeded the original commission. Troughton had no choice but to take Sir James South to court.

Upon arbitration it was decided that the work would continue to completion, at which time full payment should be made – once it was confirmed that the instrument was entirely fixed. Unfortunately South was not happy with this outcome. He halted the work and declined to pay a penny. This of course incurred further legal wrangling. South lost. But in a fit of heroic pique, he broke the entire instrument into pieces and auctioned the fragments as scrap.

Fig 3.7 The trashed telescope up for auction, with the Campden Hill observatory in the background [Image courtesy of Royal Borough of Kensington and Chelsea]

South also took pains to embarrass Troughton and all those involved in the affair by including their names in a scurrilous poster that advertised the auction. The trashed telescope and the court cases cost South $8,000! In modern terms, this was an absolute fortune.

South's pugnacity did not confine itself to Troughton. He actively opposed any road building near his observatory for fear that the vibrations of Victorian traffic would interfere with his astronomical measurements. A major thoroughfare nearby was completed only after his death in 1867!

(ONE PROBLEM THAT ASTRONOMERS USING REFLECTING TELESCOPES HAVE IS THE EXPENSIVE AND NOT INFREQUENT NEED TO RENEW THE SILVERED MIRROR SURFACES OF THIS TYPE OF INSTRUMENT.)

Dear Steve,

I live near the sea. My greatest problem is that the corrosive salty air leads to the incredibly fast deterioration of the metallic reflective coating on my 10" tele-scope mirror. A silica-protected aluminizing job lasts only three or four weeks! If I am lucky! How can I get around this problem?

Ethan
North Bend, Oregon

Ethan,

My dear fellow, don't even think of getting your mirror re-aluminized ever again. Do as I do. Spray super glue on the glass to be coated and press a pre-cut circle of cooking foil firmly on.

You may notice a slight deterioration in image quality – but at least it will be the last time you have to do it.

P.S. Please find enclosed folded piece of foil for your first attempt. No charge.

(As they say, please don't attempt this at home!)

Yes, if you sport an open-tubed reflector, resign yourself to having its mirrors resurfaced periodically. How often this is necessary is governed by local circumstances. Coastal or industrial air is bad for reflectors. Fortunately there are many facilities advertising this service. Cost is generally proportional to the mirror's surface area.

A gentle wash with warm, mild, soapy water can be used to good effect in removing aged dust, so long as the fully immersed mirror is barely touched with wipes from a large wad of cotton wool. But remember to totally douse the mirror afterwards with distilled or purified water (from a pharmacy) to remove all trace of detergent. On a clean mirror, water should flow off the surface without leaving a trace of drips.

(AS AN ARDENT OBSERVER, I CAN CONFIRM THAT WE SPEND MORE TIME WISTFULLY LOOKING AT CLOUD THAN AT ANYTHING ELSE IN THE SKY.)

Dear Steve,

The main impediment to visual observing is cloud – and up here in the north of England we get rather a lot of it. Has a way ever been found to allow astronomers to get around this problem?

Jerry
Manchester, United Kingdom

Dear Jerry,

Alas, despite the obvious need for such a thing, no dependable method has ever been discovered that can disperse cloud (at least, in practicality). A small atomic device would no doubt serve, but its after effects are rather inhibiting – an obscuring region of dust and debris, to say the least. A huge updraught of hot air (such as that from a suburban firestorm) will evaporate low-level cloud droplets. But this simply moves damp air higher into the atmosphere, where it will condense again once it meets the cooler layers above. The only real solution is to rise above it, so to speak. Mountaintop observatories succeed here precisely on this point, being above the main condensation layers.

This does not solve your rather more prosaic situation. But I do hear that help may be at hand. Telescope manufacturers, tired of competing with each other in the fields of GPS navigation systems and optics, are now working on 'Stratospheric Telescope and Astronomer Rigs.' STAR platforms are on trial that feature peripheral inflation bags similar to small weather balloons. These fill with helium supplied from a canister fitted to the telescope platform. Instrument and observer rise swiftly into the air, through the cloud ceiling and up into the sparkling clarity of the upper stratosphere. Their availability has been delayed, I understand, by various teething problems.

The main difficulty, apparently, has been the reliability of the integrated O.R.S. (Observer Return System). It has either worked too quickly – resulting in a rapid and tragically fatal descent, or not at all – leaving the hapless observer to wander the jet streams until 'rescued' by a military missile. In fact, it is rumored that a recent plane disaster was caused by its impact with a STAR test pilot. We must wait and see. Until then, use cloudy nights for earthly pursuits, such as reading books on astronomy. I do.

Ironically, cloud plays a big part in every astronomer's life. We spend an inordinate amount of time looking at it, fearing it, cursing it. Yet high thin cloud, only partially opaque and barely discernible, can be beneficial. It can reduce glare for planetary observations and even act as a calming influence to stabilize otherwise erratic "image-shaking" air currents. Low thin cloud is reputed to have allowed sunspot observations by medieval astronomers – but I would not recommend it. Indeed, NEVER look directly at the Sun – it could be the last thing you ever see.

The nature of different cloud forms can betray how the sky is shaping up. It is worth reading up on these characteristics. For instance, if you notice that stratus is developing late in the day, plan something other than astronomy for the evening. However, if cirrus develops soon after the passage of a low pressure weather system, then ready the 'scope for a full night's steady air observation!

Fig 3.8 Venus – definitely NOT a planetary surface from which to observe the universe! [Image obtained by *Pioneer* 1979, courtesy of NASA.images.org.]

(CLOUD IS NOT THE ONLY ENEMY OF OBSERVERS AND THEIR TOOLS.)

Dear Steve,

Like many others in the southeast of England, my property was blasted by hurricane force winds on the night of October 15, 1987. Although I was not, of course, observing at the time, I was using the cloudy conditions as an opportunity to carry out useful maintenance inside my garden dome, which houses my 14" reflector. I heard the wind howling outside but carried on working away.

Indeed, I was so engrossed in a lengthy overhaul of my sidereal drive that it was not until 2 a.m. that the screaming crescendo of wind-power outside began to alarm me. I stepped outside to see what was happening. That, of course, was my mistake. Before I knew it, a powerful gust gained entry through my observatory door. There was an astounding roar and the sound of splintering wood. Then, as one piece, my observatory rose spinning into the air – just like the farmhouse in the opening sequence of The Wizard of Oz! *It sailed into the air over the rooftops and soon went out of sight. You can imagine the shock – there we stood, my telescope and I, suddenly exposed to the ferocious buffeting of 70 mph winds. I gasped for air and strained against the roaring wind for as long as I could, hoping to spot the structure drop out of the cloud deck further down wind. But it simply vanished. I still have no idea where, or even if, it came down. My home being near the coast and taking the storm's course into consideration, I fear it may have finished up in the sea.*

My problem is this. No trace of my observatory was left after the storm. Not a stick, nut, or bolt. The circular concrete floor it enclosed now simply looks like a dais upon which I have placed my telescope mount. (Dare I add that the telescope itself, exposed to the elements and ripped from the mounting, also disappeared that night?) Consequently, my insurance company is disputing my claim for a substantial observatory building and instrumentation of which there is now no evidence.

I understand that you frequently drop in on various amateur astronomers to admire their observatories – although sadly not mine, as yet. I was wondering, as one astronomer to another, if you could possibly drop me a line as if you had indeed been on a visit recently. I leave it to you as to how you describe the excellent handiwork – the oak paneling with the silver inlaid motifs, the framed signed portraits of the Apollo 11 *astronauts, the five or six laptops, the antique astronomy books, etc., not to mention the complete set of Magnus wide-field ED glass eyepieces. None of my neighbors will do this for me, as they are (I'm told) "Ruddy-well glad to see the back of the eyesore."*

Richard
Haywards Heath, United Kingdom

Dear Richard,

First, my sympathies on the loss of your observatory. Such an event is bound to be quite traumatic. However, seeing the address on your letter, I have to tell you that unbeknownst to you I have indeed visited your home and seen your observatory. You were not at home, but your lovely lady wife let me in one day when I was in the area. It appears she did not inform you of my visit. Just an oversight, I am sure. But rest assured she looked after me with great care and attention – for about five hours, I think. A splendid hostess, I must say.

But here comes the rub. For a short while during my visit (I think your wife went to the bathroom) I did indeed take a look at your observatory. Strangely, my recollection of it does not accord with your lavish description. I seem to remember a ramshackle square shack made of rotting fence panels roughly nailed together. The chicken wire and tarpaulin roof was simply a lift-off lid covering the whole affair, wasn't it? I would not refer to that as a dome, would you? I do recall some torn posters inside attached to the damp walls. I do remember those. But I am pretty sure that none of the *Apollo 11* astronauts have ever disported themselves in the manner of the figures represented by these pictures. As to the books and eyepieces, perhaps you had added these subsequently, since I observed no shelves upon which these things could have been supported during my short 'tour' – a splinter from which I am still trying to extricate. Dare I ask if you are asking me to, shall we say, embellish the value of your establishment? That, of course, would be totally abhorrent to my moral sensibilities – so I regrettably must decline your invitation to assist. Do, however, pass on my very warmest wishes to your wife.

Wind. It has been my mortal enemy on a number of occasions (preventing observations being the least of them) – and in one memorable incident was responsible for emptying a whole tin of boiling ravioli into my lap. The less said about that, the better. We move swiftly on.

Storms are not benign beasts – and can be positively dangerous for large open-air structures such as radio telescopes. During a previous UK storm, it was the giant internationally renowned 300 ft radio telescope at Jodrell Bank (the Lovell Telescope), near Manchester, England, which found itself in danger. It had only 5 years previously benefitted from a thorough refit, to strengthen a structure creaking from metal fatigue. In particular, two circular wheel girder supports were added to provide additional strength to the reflecting surface.

However, the 90 mph winds that buffeted the new metal latticework on January 2, 1976, tested it to its limits. It was during this storm that engineers noticed, with cold-sweat alarm, that the whole 3,200-metric ton edifice was being steadily blown off its azimuth driving track – and was apparently within half an inch of coming off completely when, with a flourish of genius, the telescope operators rotated it 180 degrees, forcing the storm itself to blow the telescope back from the brink the other way. Hats off to them for this remarkable quick thinking! Although the telescope narrowly survived a gust of 94 mph, two diagonal bracing struts were subsequently added at the base of the tower trunions. And when the weathermen warn of approaching storms (which they do occasionally forecast correctly), a precautionary

Fig 3.9 The Lovell Radio Telescope at Jodrell Bank, Manchester, England. Almost a martyr to wind [Image by the author]

parking procedure now aligns these towers to the wind and the dish is directed at the zenith to reduce its "weather-beaten" profile.

(WE SHARE THIS PLANET WITH MANY WONDROUS CREATURES. BUT SOME SHOULD HAVE BEEN LEFT AT THE PLANNING STAGE.)

Dear Steve,

My telescope mount and sidereal tracking system were recently colonized by an unsuspected ants' nest, resulting in a dramatic and catastrophic midnight failure of my electrical circuits. Although the insects and their earthy superstructure have now been removed, I remain vigilant to a re-invasion. Is there a way that I can prevent a reoccurrence?

Steve
North Weald, United Kingdom

Dear Steve,

The appropriation of astronomical equipment by foreign bodies is far more common than realized. And eviction is often not a simple matter, either. I well remember when my wooden-walled 15-inch reflector tube was taken over by a bat colony.

I absent-mindedly left the focus rack drawtube open one night, and by morning about three hundred of the little blighters were sleeping inside the tube. Unfortunately, being a protected species, they could not be dealt with as directly as one would wish. It remained a problem until I had the idea of hanging up one of my old canvas tents in a nearby tree. On their return from a night's feeding the next dawn, this cavernous structure was so overpoweringly attractive to their cave-dwelling instincts that they took up residence there and then, never to return to the reflector tube. It took some time to restore my instrument to functionality, however, since by this time the guano on my parabolic primary mirror was 5 inches thick. Digging that out took three ammonia-stenched eye-watering days....and the mirror had to be re-silvered! By the following spring the bats had moved on. And, after arranging a thorough six-week dunking in the solar neutrino detector tank at the Homestake mine, South Dakota (filled with dry cleaning fluid), the tent's as good as new, too. By strange coincidence, the neutrino research team discovered seven new types of subatomic particles during the same period!

Your solution is happily simple. Prevention is better than cure. Ensure no queen ants enter your 'scope's mounting alive by introducing an arachnid defense force. One adult spider per square foot of surface will be about right. They will devour those plump juicy home-seeking queen ants with relish.

(Editor's note: We understand that, shortly after this exchange, the correspondent was found one morning as a desiccated husk enveloped in silk at the foot of his telescope – mouth agape in a last gasp of terror. We therefore stress the importance of using common benign spiders, rather than employing the excessively dimensioned South American varieties apparently used in this case.)

Despite protection from the weather, equipment left standing in the backyard merely provides another desirable residence for invertebrates. Innocently switching on the sidereal drive of my f/14 5″ refractor one night, the circuitry exploded in a deafening shower of sparks. There followed the ominous aroma of burnt wiring. Investigating the problem the next day, I found that the hollow 5-ft telescope pillar (through which the electrics are routed) had been carefully filled to the brim with earth by a very busy colony of ants. Although admirably supported by three outrigging spars, my engineering mistake had been to leave the base of the slightly elevated hollow pillar open to the lawn beneath. Nesting directly underneath, the mandibular minions had steadily bridged the 3-inch ground clearance with their mounded city, then simply worked their way up.

To evict them from their skyscraper empire, I had to disassemble the entire mounting (including removal of the equatorial head) and use a broom handle to dislodge the tightly packed cylinder of dirt and animated chitin that had engulfed the wiring and main supply junction box. Replacing the charred wiring and careful cleaning took me a whole day, not including the tiresome polar re-alignment of the reinstalled equatorial head the following night. Most startling was my discovery that the antish queen had seen fit to use the entombed 240 v junction box as a nursery! Its casing was packed solid with a yellow-white froth of ant eggs and larvae. My goldfish loved them.

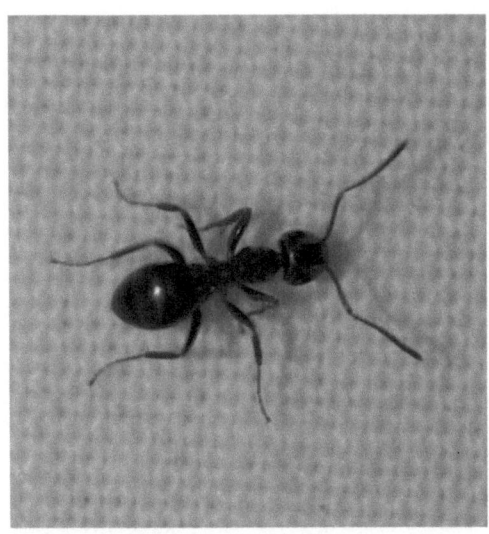

Fig 3.10 The enemy [Image by the author]

Yet even professional astronomers have problems of the invertebrate kind. Many years ago, so the story goes, astronomers at the Mullard Radio Astronomy Observatory in Cambridge, England, were having problems with the 5-km array (the Ryle Telescope). Comprised of eight 13-m dishes, these operated automatically under computer control, aiming at their pre-designated targets without human intervention.

Arriving one morning to access the results from an observing run, astronomers were astonished to discover the radio dishes pointing in all the wrong directions. Thorough examination revealed no faulty equipment. With not a little head-scratching the next session was planned and initiated. There was consternation the next morning when yet again the radio dish array was found to be in a state of. . .er..disarray. There were serious speculations that a saboteur was at work. So, the next night, an observer was secreted in the shadows of the control room to catch whoever was responsible. Late into the night, quite suddenly, the radio dishes began swinging haphazardly out of alignment. But no one was there. Yet, despite the lack of a malign human, the culprit was found!

Computerized instructions to the telescopes were being supplied by a perforated paper tape – one of the primary input devices for computers in those ancient days. The computer would read this tape as it passed between a small bulb and a light sensor. In a darkened environment, this lonely light was irresistible to a small moth that had strayed into the control room. During the night, the insect had been fluttering beneath the light as the paper tape was being read, thus imparting its own instructions for the observation of the universe.

One trusts that the moth had no squishy fate and was gently evicted. A cover was hastily constructed for the paper-tape reader and harmonious observations ensued. No journal, however, knowingly received the results of the moth's research ambitions or observations.

Fig 3.11 The Ryle 5-km radio array, Cambridge, UK. They all turned away with embarrassment as I took this picture. Hopefully the instrument is under human control these days [Image by author]

(IN EXTREME CASES, THE JOYFUL ACQUISITION OF A NEW INSTRUMENT MAY BE TINGED WITH GUILT.)

By email
From: Mary Harvey[harvem@globular.com]
To: Dear Steve[doctorsteve@help.com]
Subject: Moving on

I shall get straight to the point. I am having a wonderful time with my new 4″ reflector. The trouble is I think that my binoculars suspect something, although I have done my best to hide it from them. I have had my binoculars for five years and still have a great deal of affection for them, so the last thing I want to do is hurt their feelings. What on Earth shall I do?

From: doctorsteve@help.com
To: Mary Harvey[harvem@globular.com]
Subject: Moving on

Hi, Mary!

It appears that, having been satisfied with the binoculars for a few years, you have gotten tired with their performance and naturally have gone on to bigger things. Try not to feel so guilty, as this sort of progression is only natural. Explain to your binoculars that you still love them, but they can no longer satisfy your expanding needs. It may help a great deal if you can find someone else who can take your place in their life. Perhaps a novice in astronomy who has no equipment? It will be a trying time for you both. I wish you luck with your endeavors with the 4″.

Binoculars are often portrayed as a mere steppingstone to "proper" astronomical instruments such as telescopes. Nothing could be further from the truth. Binoculars are fully paid up and qualified members of the astronomical arsenal. They are easy to carry around, intuitive to use, and have a light grasp capability that allows proper observation of many classes of celestial phenomena. The Moon, star clusters, nebulae, and comets all fall within their remit. It is certainly possible to carry out professional work on the brighter variable stars. You can even use them in the daytime, if you want to. For example, binoculars fitted with safe filters make them dedicated daytime solar instruments.

Warming to my theme in answer to this question, I have to say that you can never have too many instruments. Keep them all. You will eventually find a use for each of them. They will all be particularly mobile, powerful, enviable, or interesting in their own right. For example, binoculars are best for wide views of a comet's tail. But for close-ups of its nucleus, wheel out the planetary refractor or catadioptric.

I have also found that keeping old equipment shows its worth in the end. Cannibalization of old equipment has often been used to vastly improve new equipment. The genetic line continues, as it were!

(IN ORDER TO AVOID HIGH COSTS, AMATEUR ASTRONOMERS HAVE TO BE VERY RESOURCEFUL IN THEIR ACQUISITION AND USE OF RAW MATERIALS FOR TELESCOPE CONSTRUCTION. IT IS A TRADITION OF LONG STANDING. BUT IT CAN BE TAKEN TO EXTREMES.)

Dear Steve,

I am, as my father before me, an avid maker of homemade telescopes. I am forever conscious of having to maintain the Olympian standards of my late lamented dad, whose inventive transformation of everyday objects into instruments suitable for high-level astronomical research was legendary. Indeed, the reputation of our family for inventive spare part construction goes much further back. You may have heard of my infamous medical ancestor, Dr. Victor Frankenstein, who made good use of his creative assembly abilities in the early 1800s. However, I have been troubled of late by our old family retainer, Igor. He came to the 'States with the family when they had to leave Europe..er. . .in somewhat of a hurry – owing to some unfortunate misunderstanding. I have to say that he is wearing remarkably well, as he must be approaching 220 years old.

But I fear that his years are at last catching up with him. I have caught him occasionally responding to me with the words "Yeth, Baron. . ." despite the fact that our family thought it prudent to hide our noble origins a long time ago. However, that is not the worst of it. Having served me well over the past few years with supplying raw materials for my telescopes, he seems to be reverting to type.

Piece by piece cherished articles, which have hitherto formed the loving fabric of my house, have unbeknownst to me been taken into my workshop at dead of night by Igor to be frenziedly transmogrified into yet another observational aid. Each morning, I have woken to find more furniture and fittings have disappeared, with only a few traces of sawdust as mute testament to their passing. In the workshop, I find that yet another unfamiliar 'scope or accessory stands newly glistening in a fresh coat of varnish. But lately it has gotten worse, as he seems to have moved on from commandeering merely inanimate objects. Only last week, our son's pet rat disappeared one night. In dread, I searched in the workshop. Can you imagine my horror when I discovered that my eight-inch reflector now boasted a new furry end cap, four tiny claws serving as tube clips? And a little furry head as an eyepiece cover. I questioned him and bitterly complained to him about his cruel use of the animal. But he simply grinned his twisted mouth, grunted, and said "..yeth, Baron."

Following that, he made a great burglar alarm using a nearby coyote's howl. But I have not heard that coyote since.

Even then, the potential problem did not really sink in. It was this morning that set me writing to you in a last forlorn scream for help. For I awoke to find our neighbors missing. I haven't been down to the shed yet. Dewy overnight foot-prints in the grass reveal evidence that someone has been there. Shreds of torn clothing hanging from the shed door indicate that a horrific but silent struggle has taken place. I can see spatters of dried blood on the inside of the shed window. The chain saw is missing from the under-stairs tool cupboard.

I am in dreadful torment. One side of me reels in abject terror at what Igor may have done; the other is eager to see what fantastic piece of grisly equipment he has produced.

Victor Jnr.
Pontiac, Michigan

Look my good fellow – you really must get a grip on yourself! This practice has a long history. The concrete pillars supporting professional tele-scopes all over the world are simply stuffed full of human remains – often intentionally. So, please, do go down to the bottom of your garden to see what awaits your pleasure in the workshop. I'm sure you're in for a pleasant surprise. (Just in time for the Jupiter opposition, too.)

The ATM (Amateur Telescope Maker) tradition is alive and well in astronomical circles. In the home of any amateur astronomer, there is a constant danger that household goods will be appropriated for a purpose they were never originally designed for. The tubular cardboard inserts of toilet rolls and new carpets are obvious examples. This art form is best exemplified in the publication "Make Your Own Telescope from Everyday Materials" by Reg Spry – South Downs Planetarium Trust, 1998, ISBN 0-9531-716-04, 48 pp., £4.95 (paperback).

As the dimensions of telescope optics available to amateurs have become larger, the cost of sufficiently powerful equatorial pivots to mount them has increased exponentially. American Frank Dobson had the brilliant concept that if you used these big apertures for their excellent light grasp only – at low magnifications – a simple up-down axial movement on a rotating base plate would suffice. And they could be made simply from readily available materials found in the home. His invention may be entirely responsible for the survival of hardware superstores.

And yes, human remains have indeed been used in telescope construction; at least, in the foundations of one. James Lick was a wealthy nineteenth-century landowner who, at the age of 77, had a stroke. Perceiving an impending demise, he gave some thought to a fitting mausoleum. He gave serious consideration to a giant pyramid in San Francisco! Fortunately for posterity, San Francisco and the philanthropic reputation of Lick himself, he was persuaded in the end to bequest (what would be) the biggest telescope in the world. We may have influential friend and prominent Californian scientist George Davidson to thank for guiding Lick towards this more sensible legacy. Thus was initiated the Lick Observatory atop Mt. Hamilton. Lick died three years later in 1876, occasioning, it was said, one of the most fabulous funerals ever seen. Yet his final request had yet to be played out. In 1887, as the observatory's foundations were being completed, his body was reinterred at the base of the pier that subsequently supported the giant 36-inch refractor – where it remains to this day. Now that's what I call being buried in style.

The most curious "scrap" appropriation I have indulged in is to wrest silk from a live spider to make a finder scope's reticule!

(AT AMATEUR LEVELS IT IS INCREASINGLY IMPORTANT TO SQUEEZE EVERY OUNCE OF CAPABILITY FROM A TELESCOPE. SOME ENTHUSIASTS ARE QUITE INVENTIVE IN DOING SO. . .)

Dear Steve,

Toying with a few lenses last month, I quite by chance found that placing a couple of them at a certain location in my telescope vastly increased what I could see with it. Whereas before I had trouble seeing the belts of Jupiter, I could make out the volcanoes on Io! Thrilled to bits, as you can imagine, I went straightaway to my friend John and showed him my new invention. He, too, was impressed. In fact so impressed that he asked me to lend them to him for a while, which being flattered of course I did.

You can imagine my shock when he started showing off his new observations to the others in our astronomy group, made with what he now calls HIS invention. He denies my involvement completely. What should I do?

Dustin
Sioux Falls, South Dakota

Dear Dustin,

Tricky. Since you obviously had no time to take out a patent it will be difficult to prove anything. I fear that your only option is sabotage.

I gather that since you state you were playing around with lenses that you have quite a number of them. Somehow (and for the purposes of avoiding the charge of incitement to commit a crime we shall not define) you must deftly swap a component of the special device with another of similar appearance. You then ask your 'friend' John, in a loud voice, why the device is no longer doing its job.

Having not devised it in the first place and therefore not aware of how it DID work, he will now be hard put to explain why it no longer does so! You grandly enter the scene and then, with great proclamations of functional elaboration, place the correct lens in its rightful place and stand back to general applause. Where you now put the spare lens is up to you, but if your friend is still nearby?

Innovations should be spread far and wide for the benefit of us all. But it is only fair that credit should go where it is due. Consider the woeful tale of Chester Moor Hall.

Having cornered the market on reflecting telescopes, Isaac Newton's pronouncement in the late 1660s that refractors would never be achromatic was a brilliant marketing ploy! Yet only a little later, in 1695, David Gregory, nephew of the inventor of the Gregorian telescope, suggested that chromatic aberration (false color) might be overcome by a clever combination of lenses.

About 30 years later Chester Moor Hall came across this "throw away" line and in 1733 designed a refractor lens made by combining two glass elements of differing refraction. With no practical skill of his own he subcontracted their construction. Being mindful to keep it secret, he took pains that the same

London optician did not work on both glasses. He therefore commissioned one element to be done by Edward Scarlett of Soho and the other by James Mann at Ludgate.

Both companies were very busy with other work, and they in their turn subcontracted – to the very same firm of George Bass at Bridwell! Noticing that each glass of the two separate commissions was the same diameter, Bass put these together and the secret was out – at least to George Bass. Chester Moor Hall kept the solution to himself and strangely so did George Bass. Twenty years later, Bass mentioned the new design to Peter, the son of telescope maker John Dollond. The upshot was that John Dollond ran with it to the patent office as fast as his legs could carry him, despite (so the story goes) having nothing to do with the original development. There ensued a bit of a fuss – including a lawsuit brought by the now aggrieved Chester Moor Hall. The judge decided in John Dollond's favor, decreeing that those who make the invention public should profit by it. The rest, as they say, is history.

John Dollond and his son Peter built up a world-famous telescope-making firm whose optics directly advanced astronomical research. The modern-day descendant of this company has now reached the dizzying heights of being a UK high street optician. Oh, and of Chester Moor Hall? It apparently says on his tombstone that he was an excellent lawyer and mathematician. His expertise does not seem to have extended to good business sense, however.

(THERE CAN BE NOTHING MORE EXCRUCIATING THAN THE ANGUISH AN ASTRONOMER SUFFERS BEFORE RECEIPT OF AN ORDERED TELESCOPE.)

Dear Steve,

In April of 2004, I ordered a simple telescope kit from Scopes-4-Dopes, the company that promises an uncomplicated telescope construction kit. They said it would be with me soon. In May 2004, they sent me the tube. Just the tube. In the following months, they sent me either a screw, washer, bolt, or nut. All I have for my money at the moment is a tin tube and a bag of mixed 'shrapnel.' I have just phoned them and apparently had some confusion cleared up. They now say that it was always made clear that a single component would be sent regularly, building up month by month, until I had all the pieces I needed to finish the job. When I asked when I would receive my final part, he said probably in the summer of 2035, although he was still trying to locate a source for some of the later bits. (I think I heard someone laugh in the background at that point.) Do I wait for the rest of it, or cut my losses and get a ready-made one from someone else?

Luke
Wausau, Wisconsin

Dear Luke,

Oh dear! It is fortunate indeed that you have written to me, for I have only just completed one of their kits – and I started mine in 1949! And believe me it's not worth it! A heap of junk. It did not help that the connector bolts went from being imperial to metric threads halfway through, either. If you can, send the bits back and get a refund. If not you can have mine – and if you can work out where part 97a goes, you're a better man than me!

I suppose there are some of you out there that might scoff that telescopes could take so long to construct. Fact, as they say, is often stranger than fiction. Take the Radcliffe Telescope in Pretoria, for instance.

In 1924, Harold Knox-Shaw of the Radcliffe Observatory in Oxford, England, expressed the need for a large southern hemisphere telescope to augment their own rather dated instrument. Just a few years later, in 1929, the great car magnate Sir William Morris (later dubbed Lord Nutfield) charitably donated 100,000 GBP to purchase the old Radcliffe Observatory and grounds for the expansion of the Radcliffe infirmary. Armed with these funds, the Radcliffe Trust soon found a site near Pretoria, South Africa. No less than the famous amateur astronomer W. H. Stevenson was dispatched to spend 6 months there to check the location's suitability. (Nice work if you can get it!) In 1931 the old Radcliffe Observatory was sold and preparations were made for the new. The UK Attorney General objected, however, to charitable funds being spent abroad. Oxford University also protested, believing the money should be theirs.

There ensued an amazing star-studded legal action in which luminaries such as Albert Einstein, Sir Frank Dyson, Arthur Eddington, and Harlow Shapley testified. The Radcliffe Trust obtained a narrow victory in which they were allowed to use just 65,000 GBP of the pot!

Fig 3.12 The court case involved Einstein – seen here in typical self-deprecating mode for the benefit of a passing photographer

In 1935, a 74″ mirror blank was ordered from Corning Glass in the United States for an expected July 1936 delivery at Grubb-Parsons, Newcastle (UK) for grinding and polishing. Unfortunately Grubb-Parsons soon began receiving apologies for the blank's delay. The catalog of woes included a damaged furnace, bad weather, floods, volume of other work, contaminated glass, and annealing problems. In all it took three castings. As a result, the mirror blank was not shipped to the UK until October 1938, 2 years late.

Meanwhile, the telescope superstructure, built in the UK then disassembled for transport, had been shipped to Pretoria. Re-assembly at the other end of the journey had to wait until a truckload of parts being driven to the site was tracked down after getting lost. Not that it helped much, because on arrival it was found that parts of the turret dome had been damaged in transit and had to be forcibly bent back into shape. Engineers then found that without a crane big enough to lift it through the dome slit, the 15-ton polar axis could only be fitted if they smashed a hole through part of the newly completed observatory floor. The main electrician then disappeared.

In February 1939, the chief optician working on the mirror in Newcastle dropped dead on the job, and his replacement had problems grinding out an imperfection in the mirror's figure. On September 3 Britain entered World War II, and work on the mirror was suspended. Fearing possible bomb damage to the fragile glass, it was suggested that it be buried for the duration of the war. Aghast, there followed a command from the Radcliffe trust in Oxford that the mirror be sent to the safety of the United States for completion. But this correspondence arrived too late; the mirror had already been buried. The onset of winter froze the mirror into the ground, and everyone was forced to wait for the thaw. Work recommenced, but not until October 1947 was the mirror completed.

The mirror arrived in Pretoria in May 1948, having been delayed by a gale and its ship blown off course. On mounting the mirror into the telescope superstructure it was found in tests that the stellar images suffered from severe astigmatism. It had taken so long for the mirror to arrive that all of the mirror support pivots had rusted solid. Rectified by a can of oil, the telescope was up and running by December 1952, over 22 years after the project was initiated!

Fig 3.13 The Radcliffe Telescope [Image courtesy of the South African Astronomical Observatory]

(SOME PROBLEMS ARE JUST YOUR OWN FAULT.)

Dear Steve,

You've just got to help me! I'm desperate. I need you to tell me what I've got to do. But I guess I'd better begin at the beginning.

I'm an amateur astronomer. Like many young fathers living in Florida, I was really looking forward to my young son becoming old enough to appreciate a trip across state to the Kennedy Space Center. Oh boy, we were going to have such a great time. Well, he was ten this year so we finally made the journey. But what's happened is just more embarrassment than a proud dad should have to bear. Don't get me wrong, we had a fantastic day. We got up close and personal to the rockets and spacecraft, we did the shows, got to talk and eat with a real astronaut, touched a Moon rock — and we flew three times in the shuttle launch simulator. Then it happened.

'Course, once we'd covered everything on the site we made a bee-line for the Space Center Gift Shop. What a mine of goodies! Model spacecraft, NASA logo clothing, memorabilia and posters everywhere. But suddenly for me one thing shone out more brightly than anything else. A space helmet. OK, I knew it was for the kiddies — just from its size alone. But putting on a spaceman's helmet was something I had dreamed about since I was my son's age. Well, somehow, I squeezed it on. That hurt. But I looked in the mirror nearby and I looked fantastic. Maybe the big red and blue voice-changer knobs on the side were a bit of a giveaway — but there I was, astronaut of the fleet. I was in heaven. Until I tried to take it off.

I just couldn't get the neck ring past my ears. I swiveled it around and around, bent my head, smothered my neck with soap from the washroom — but it didn't budge. My noisy desperation began to cause a scene. My son got upset. Others in the shop began to crowd around, some laughing, others scowling — believing I had stolen it off my son (who was by that time crying). For a while I couldn't even see 'cause my heavy breathing misted up the visor — until I found the release clip that flipped it up. Eventually, my antics caused so much disruption that I was asked to leave. Of course, I had to pay for the damned thing on the way out, too! My son was really miserable, as he hadn't gotten anything from the shop. It didn't end there. On our way back to the car people kept stopping me for autographs — thinking I was a real astronaut doing a walkabout! My son was inconsolable and just consumed with shame.

Well, here I am, writing to you. It's still on my head. If it hadn't been for the hinged visor I'd be dead from hunger or asphyxiation by now. I can't destroy it to get it off 'cause my son, having left the Space Center with nothing, says he wants it. How can I refuse him? It's been a week now, and you can imagine what I'm going through – especially from the guys at the office. They even found the little button on the back that plays the stars and stripes. That can get really irritating, you know? Don't suppose you have any suggestions, huh?

Deke
Miami, Florida

Dear Deke,

Sorry about the delay of my reply. I've been admiring the picture you sent me. It is sort of cute, isn't it? I just love the little bobbled antenna sticking out at the top. No, seriously, let's get down to business. I can see your dilemma. Wanting to keep it in one piece is a bit of a sticking point (tee hee!). Sorry – I'll calm down now. There seems only one option. What you may not know is that the size difference between a child's head and that of an adult is not as large as you might think. The young head the helmet is designed for is quite disproportionately large. Your head, believe it or not, has acquired quite a lot of surplus material on the outside during maturity. (What's been happening on the inside is a matter of debate.) Therefore, your only answer I'm afraid lies in a little starvation; not to a deadly degree (that would be silly), but sufficient to remove all that ugly fatty stuff above your neck. Three weeks on meager rations should do it. Give it, say, to the end of the month; then make a determined effort with generous helpings of soap and axle grease. Good luck!

Understandably, in space, it is not a surfeit but the possible lack of a space helmet that occupies an astronaut's thoughts. An unsecured perhaps faulty neck ring or glove seal, a stray diminutive meteorite or even a misjudged swing with a tool – a major breach in a spacesuit is always possible, although hopefully not very likely. Nevertheless, there has always been much morbid speculation about what would happen to our hapless spacewalking hero in these circumstances.

Down here on the surface of Earth, the weight of the atmosphere around us exerts a constant pressure against our bodies, equivalent to about 15 lb/inch^2. The perambulating sack of watery material that is a human has evolved in adaption to this pressure. This pressure is carefully reproduced when we leave Earth to lark about in spacecraft and spacesuits. Our blood stays liquid because it is sufficiently compressed by this surrounding air pressure. Without it, blood would turn to gas. It would be like the bends, that condition where nitrogen unnaturally dissolved in a deep-sea diver's blood returns to gas once the pressure of the air falls back to normal. Reduce the pressure even further to zero, in a catastrophic exposure to the vacuum of space, and at a fiery 37°C all the liquid within the body can boil – and does so. Without an atmosphere hemming things in, the normally benign countering pressure within the body will explosively expand internal cavities and soft tissue – squeezing the liquefied remains of the unfortunate astronaut through the unforeseen aperture like toothpaste. The last fate awaiting our luckless explorer is that, deprived of a cohesive insulated layer to seal in the heat, the "boiling" body would be almost instantly frozen. A space "temperature" of about −100°C is pretty unforgiving. One wonders therefore at what artistically frozen explosions of human protoplasm might thus be created.

(SOMETIMES YOU GET TO THE END, ONLY TO FIND YOU'VE REALLY ONLY JUST BEGUN.)

Dear Steve,

I wonder if you can help me? For twelve and a half years now, I have been designing and building a 19" reflector and a geodesic observatory in which to house it. I have even used a computer to assess all measurements and angles of its construction to ensure that the project was a success. I have incorporated the very latest of electronic technology in my 'finding and guiding' control system; and everything, from the rotating dome to the interstellar co-ordinates indicator, works absolutely perfectly. Except for one thing. My polar axis. I have discovered it's pointing in the wrong direction. In fact, instead of pointing at the north celestial pole, it is aligned to a point 90 degrees away on the southern meridian! In other words, I have the telescope mounting facing in completely the wrong direction! Trouble is, the axis superstructure is heavily encased in 5 tons of waterproof steel-reinforced hard-cored concrete. Is there any easy way out of my predicament?

Bob
Houston, Texas

Dear Bob,

In a word? No. It would seem to me that you have made an almighty cock-up of the whole thing, probably through using one of those confounded computer things. Never ever trusted them, never will!! Your only choice would seem to be to move out and start again. Explain to any prospective buyer of your house that the construction in the garden is a rather elaborate gazebo – and hope that they fall for it. (If they do, they most certainly deserve to.)

Such fundamental mistakes are easily made. You can indeed do something abysmally silly even after thinking long and hard about it. Sometimes these mistakes are a little too close to home for comfort – my local astronomical society, for instance. During the group's foundation years, when they slowly and painstakingly built an observatory to house a 16″ reflector, careful pains were taken to align the pillar support along the north/south meridian. It was not until the erection of the telescope drive atop this concrete plinth that it was realized it had been unwise to survey the meridian using the alignment of a compass – since, as everyone should know, a compass needle hardly ever points *true* north (only to its magnetic counterpart, which may differ by many degrees!). With the drive mechanism showing an obvious 9 degree offset to that of the supporting structure, that angular incongruity remained long after as mute evidence for that long forgiven error. Still worked fine, though.

But let he who casts the first stone not be living in a greenhouse (or something like that). Long, long ago, before I knew better, I once owned a precious eyepiece. Note I say "once." It was one of an identical pair I owned. They were an old ocular of American manufacture (circa WWII?) having a fine wide field Erfle design. There was only one problem. One part of the eyepiece, a cemented element of two lenses, had a blemish frozen within the thin smear of adhesive between them. I could have left it. It was almost invisible when used on a telescope – especially against a dark field. But I wanted to fix it. So I set about doing so.

Armed with a set of fine watchmaker screwdrivers and lens-ring calipers, I steadily and carefully disassembled my way towards the offending component. It was like a medical procedure. Tools were laid out on a clean handkerchief, regimented like the forceps, scalpels, and swabs you see on a stainless steel tray in the operating theater of TV hospital dramas. Except in my case these were screwdrivers, calipers, and optical cotton buds. Oh yes, don't forget the notebook, recording the order in which the eyepiece parts would be decanted from the barrel – so they could be put back in both the correct order and the right orientation. Gently, at the last, the glass mass of the infected lens was lifted out and placed on a clean cloth for examination.

The blemish was now revealed to be an area of the adhesive between the two glasses that had become discolored. It may have been simply a chemical reaction brought on by age, but I had heard, too, that Canada balsam (the adhesive used) can be attacked by fungus, of all things! What I did know is that Canada balsam melts with heat. All I had to do was cook it, separate and clean the two bits of glass, melt some *fresh* Canada balsam onto one of the pieces, then carefully squeeze the two together again. Voila! One new eyepiece!

Of course, in the years that followed, I learned of better ways to heat up glass to remove this type of glue – one of them being to gently simmer in hot water. But before then, I was far too clever. I had an infrared lamp. A one kilowatt infrared lamp. Easy, or so I thought. I placed the optical patient on an operating table of black cloth (to efficiently absorb the heat – clever, eh?) and via an angle-poise lamp-holder placed the infrared bulb less than an inch from the glass. I then stood proudly back to monitor the progress of my Gerry-rigged process. Sure enough, through the shimmering air, I soon noticed the change in the clarity in the balsam cement layer that indicated its transformation to liquid. When this change had spread across the entirety of the adhesion layer, I lifted the lamp away and took hold of the lens in order to prize its components apart. Too late, I realized my mistake.

The glass was, of course, burning hot. Incandescent. I might just as well have pulled a coal from a fire. Before I could pull the two bits of glass apart (still stubbornly held intact by surface tension) I had no choice but to let go and let fly a scream of pain. Incredibly, the still adhered component hit the table exactly on the black cloth from which I had lifted it, almost edge on. But the off-vertical angle of impact enabled the glasses to slip a little on their lubricant of soft Canada balsam – exposing crescent-shaped internal surfaces of sticky glue. As I watched, in cinematic slow-motion, these took hold of some of the dark fibers of the black cloth. In panic, lest the fibers become stuck fast, I lifted the black cloth away from the lens. The lens was glad of the unexpected additional momentum. It now rolled across the table and leaped to almost certain oblivion towards the floor. I had time to thank the gods that carpet was laid there to break the fall.

But the gods were not helping; they were laughing. The red-threaded carpet was made of artificial fiber with a fairly low melting point. As the still sizzling lens bounced its merry way across the floor, it formed around itself a growing crimson bushy cocoon that enveloped it with impressive thoroughness. It came to rest looking like a lenticular ball of wool. But I was not finished with it yet. Even then I thought I might save it. With a spare handkerchief I grabbed the steaming lozenge before it could melt a hole through my carpet (à la China Syndrome) and tossed it into an icy glass of water that just happened to be handy. It was during this rashly conceived cooling that I watched the glass components of the lens gently craze over and disintegrate. The rest of the eyepiece followed the fragmented crusty fuzz-ball into the waste basket.

Fig 3.14 My surviving identical sibling of the wrecked ocular. Colloquially known as the "Palomar" eyepiece, it was rumored to have had one of its kind employed on the eponymous telescope; given its 3.5″ diameter and 2½ lb mass, probably as a counterweight! If the rumor is true, this crusty old guy would have yielded a stunning 60 degree apparent field at ×52 on the great 200″ reflector [Image by the author]

(PRIDE OFTEN COMES BEFORE ANGST.)

Dear Steve,

It started with the setup, for God's sake. I'd been really looking forward to using my newly acquired 8″ Schmidt-Cassegrain Go-To telescope. The sky was brilliant, but it had been raining heavily for a couple of days. Bad idea. Carrying the telescope mount around my backyard looking for a firm bit of ground I slipped forwards into the mud. Down I went on top of the tripod and its fork mounting. You would not believe how painful that was. Worse, my lovely equipment had sunk almost completely into the lawn under my weight.

The tripod legs cleaned up fine under a running tap, but the fork mount motors were a different ball game. I had to clean out the disassembled electronic circuitry and gearing with cotton tips. Towards the end I had to use toothpicks and meat skewers on the drying mud! When I finished, it was still clear outside. So out I go and power up, quickly completing the star-align procedure for my Schmidt-Cass.

The motors didn't seem happy. They weren't making the same sound as last time. I set the hand control to find Jupiter, and the 'scope whizzed away and found it, a few degrees above the horizon in the southeast. What a great view! I noticed a yellowish speck to one side of the Great Red Spot, quite distinct. I'm a great one for drawing my observations, so I got busy sketching this new minor feature. Then it happened. The 'scope suddenly started slewing of its own accord, and Jupiter shot out of the field of view like a bullet! The 'scope panned up towards the zenith until it reached the gibbous Moon, where it stopped. I punched Jupiter into the hand control again, and on re-acquisition continued my sketch. I had hardly put pencil to paper when the motors sprang to life and I found myself looking at the Moon once more. The controls were clearly playing up, but I desperately wanted to record this unusual Jovian detail. So I forced the 'scope back to Jupiter. Thirty seconds later, back to the Moon it shot.

I gave up. It's driving nicely on the Moon, so what the hell, let's look at that instead. The waning terminator was cutting nicely across the Sea of Crises and further along that marvelous crater Petavius was a fantastic sight. Then the 'scope panned down to Jupiter. I lost it. Simply lost it. I screamed. Oh my God did I scream. Swearing. Jumping. Flailing. Then I stopped and noticed that the tele-scope was laughing. It was laughing at me, this maniac leaping about the garden. Its mocking laughter was deafening. I couldn't stop it, even with my hands to my ears. Then it came to me. Yes, what joy, I thought. How wonderful. I went to the toolshed. Fantastic; the axe was just where I normally leave it.

I don't remember much of what happened then. The police tell me I was still swinging the axe when they arrived. Their knockout-dart may have affected my memory. My trashed 'scope is still on the lawn, what's left of it. I can't bear to go *out there. I saved for years and look what I've done. Please Steve, I'm absolutely inconsolable. How can I bear it? What can I do? I can't just sit here sobbing all day.*

Keith
Tonbridge, United Kingdom

Dear Keith,

What can I say? I can hardly soften the blow for you. The damage appears to be irrevocable. Would the house insurance provide replacement? I don't suppose they'd look kindly on the cause of such a claim, would they? Worth trying, though – especially if you can get a written medical opinion. Do enough pieces survive for a running repair? From what you infer, I'd guess not. But you must recognize that it is more important that the psychological damage is repaired. I do recommend counseling. It helped me, years ago, when a revolving observatory dome sliced the top off my left index finger. I couldn't go near an observatory or telescope for months afterwards. My doctor called it Observaphobia. However, little by little, I was able to approach astronomical equipment without breaking into a cold sweat. Apart from an inexplicable aversion to a certain type of eyepiece I am completely cured now. Do try it, soon as you can. I've sent you details of a good consultant. He had an 8″ Schmidt-Cassegrain once, too.

The history of professional and amateur astronomy is littered with stories of astronomers being injured (sometimes seriously) by their equipment. Fortunately, accounts of equipment being harmed by astronomers are rare. I would like to relate two cases; one fairly prosaic, the other definitely not.

The first case was related to me years ago by a well respected antique telescope restorer. A man one day gingerly brought to him fragments of glass and the completely flattened tube of a rather long multi-draw Victorian refractor. It transpired that his spouse, angry at the time he spent on astronomy, had secretly placed the instrument in their driveway under the wheels of his Land Rover. Leaving for work the next morning, he had unwittingly driven over it. The enquiry, apparently, was whether it could be repaired. Looking around carefully for signs of a Candid Camera and finding none, my associate gently broke the awful truth to him.

However, in a bizarre event that occurred almost 40 years ago, we have what must rank as the most shocking attack on a piece of astronomical equipment ever. It is the tale of a telescope that was shot.

This heinous crime took place at the McDonald Observatory, Texas, one night in February of 1970. The unfortunate telescope was the 107″ (2.7 m) Harlan J. Smith reflector. A newly employed member of the staff, suffering a breakdown, carried a handgun into the dome. He fired first at his supervisor. Apparently unhappy at missing, he then took aim at the 107″ diameter primary mirror of the telescope and emptied the rest of the clip into it. It is said that the gunman attacked because he wanted to halt the telescope being used to talk to God. Although the fused silica mirror did not break, three small craters were punctured into the pristine optical surface. Incredibly, the mirror survived. The craters were carefully drilled out and painted matt black to avoid unwanted reflections. Still very much in use, the only adverse affect to the instrument was that the equivalent of a single inch of light-gathering ability was shaved off. A notorious event at the time, it was even featured on the Walter Kronkite show. Allegedly, telescopes everywhere behaved impeccably for months afterwards!

(ASTRONOMICAL RESEARCH ENFOLDS ALMOST ALL THE OTHER PHYSICAL SCIENCES. AS SUCH, THE INSTRUMENTATION USED TO PLUMB THE VAST UNKNOWN REACHES OF SPACE CAN BE EXOTIC OR BANAL. SOME OF THESE DEVICES OF EXPLORATION MAY BE VERY COMMONPLACE INDEED.)

Hi Steve!

I understand that you wear spectacles. As an astronomer who is known as a proficient visual observer, does your failing eyesight worry you? I love using telescopes to gaze at the fantastic things that they show us. But I have just discovered that I, too, now need glasses. Where do I go from here?

Catherine
Casper, Wyoming

Dear Catherine,

It really depends on why you need them. If you need correction for long or short sight, you can still observe without glasses – it's simply a matter of racking the focus in or out. Astigmatism is a little trickier. But did you know that some eyepiece manufacturers even incorporate lens mechanisms that can cater to this, too? In my case, it was only one eye (my preferred observing eye, damn it) that gave me problems – so it was simply a matter of training up the remaining good eye for astronomical observations. Since you do not go into detail it is difficult to assist you more.

However, even if the problems with your vision are insurmountable, all is not lost. If your days of photon reception are truly over, move on to detecting another type of particle – the graviton. Admittedly this is a new, yet rapidly developing, field in the realm of amateur observation. But be encouraged that, as yet, even well financed professional gravity wave researchers have failed to detect a single graviton. This is why we amateurs are stepping up to the plate.

Gravitational principles provide us with a head start by determining that success is governed by the mass of the signal receiver. Professional institutions have approached this by constructing huge heavy weights delicately suspended from sensitive detectors that are capable of sensing the infinitesimal motion that a passing gravity wave will impart. As always, the amateur approach, though radical, is delightfully simple. We become the detector ourselves!

Becoming large enough for the purpose obviously involves a great deal of eating – so you can see there are already wonderful fringe benefits in this field of research. Dedication is required, in order for your body to reach the great weight required – continuous consumption of pies, pastas, ice cream, and cakes is absolutely necessary. You can see how easy it is to warm to the task!

Basically what will happen is this. A nearby gravitational event – such as a stellar implosion, neutron star collision, or even, in theory, a heavy man slipping on a banana peel over three blocks away – will create a gravity wave traveling through space towards you. This density pulse, though weak, should generate a perceptible wobble in your grossly distended stomach – possibly otherwise interpreted as a mild case of indigestion. Even the relative strength or proximity of the gravity wave can be determined. A distant, weak gravity wave will merely result in sufficient discomfort to cause a wayward cake crumb resting on the crest of your stomach to be dislodged a tenth of a millimeter. A small effect, but well able to be noticed by a willing, watchful scientific collaborator. Very strong gravitational waves will be energetic enough to generate a bowel movement. You see how exciting this field of amateur research may become.

In your gluttonous preparations, do be careful not to introduce flatulence, as this can disturb the signal-to-noise ratio. So, sell your stupid old telescopes and start eating for science. Feel sorry for the skinny visual astronomers. They will have to stick to measly old photons!

Gravity waves were predicted by Einstein's 1916 work on general relativity, falling out of the assumption that all forces in the universe (including gravity) are mediated this way. Electromagnetic radiations are communicated by photons; the weak nuclear force has "W" and "Z" bosons do its work, and the delightfully coined gluons are required to exert the strong nuclear effect. It at first seems sensible to assume that gravity (another force of nature that exerts an effect at a distance) uses a particle to communicate its force. But despite the best minds being set to the task, gravity refuses to be quantisized and thus allow a search for "gravitons". We are therefore currently forced down the route of wave detection. Unfortunately, although paradoxically it has the highest effect on what happens to the universe at the macrocosmic level, gravity is fundamentally the weakest of the four "forces" of nature by several orders of magnitude (preceded by the electromagnetic, weak, and strong nuclear). This means that you either need a very powerful stellar or cosmological event for a gravity wave to be produced or an equally massive or sensitive detector.

Detectors past, present, and proposed generally rely on the principle that a passing gravitational wave should impart a proportional "wobble," either in a large object or in the actual space between objects. Dear Steve's aforesaid biological suggestion is therefore well founded in the current science – and I see around me that some keen observers have already taken me at my word! However, one problem that has dogged this field of research is differentiating a signal between that generated by a nearby stellar cataclysm and the heavy nearby tread of students heading for the college cafeteria. It's the old signal-to-noise ratio problem – and one that is fraught in gravitation because the signal is so intrinsically weak. Nevertheless, many scientists have taken up the challenge.

The first deliberately designed gravity wave detector was that of Joseph Weber's at the University of Maryland. Weber tried to detect gravitational resonances in a heavy bar of aluminum weighing 3.5 metric tons, but his results could not be duplicated by other scientists using similar designs. Other designers have employed the trick of interferometry using the multiple reflection of light from accurately placed mirrors to catch the brief change in distance between them caused by a gravity wave. Of course, the greater the distance between reflectors, the more sensitive the detector. In a project begun in the 1990s, known by the name LIGO (Laser Interferometer Gravitational-wave Observatory), Caltech scientists used an "L"-shaped detector that had arms 4 km long. But even at these dimensions, the expected displacement due to the careless passage of a gravitational wave may be only one hundred millionth the diameter of a hydrogen atom. I can imagine that a butterfly impacting their laboratory wall would elicit a far greater response. This is why they operate a pair of these devices 2,000 miles apart — to isolate local causes. The chances of two butterflies hitting the wall at the same time in both locations are . . . .um. . ..

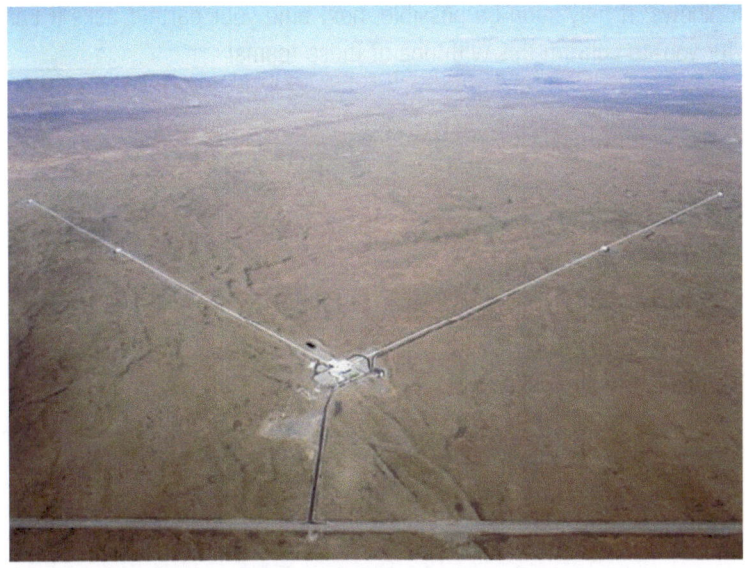

Fig 3.15 Caltech's LIGO project; searching for gravity [Image courtesy of NASA.images.org.]

The VIRGO project, a collaboration between Italian and French scientists, uses much the same principle, with two arms 3 km long. However, devious use of mirrors in multiple reflections create an effective total length of 120 km! Their boast is that they "should" be able to detect rumblings from as far away as the Virgo cluster of galaxies. Sure. Not to be outdone, elsewhere in Europe we have a German/UK team using the GEO600, which although sounding like this season's new lawnmower is actually another interferometer — this time using a laser beam tunneling through two vacuum tubes only 600 m long. Recently upgraded, the claim for this highly vibration-damped device is that it will sense gravity ripples left behind by the Big Bang. Hopefully they mean the one right at the beginning, not the one caused by the slamming of the door at the betting company Ladbrokes – who were running a 500-1 book on the detection of gravitons – but in light of the sophistication of latest attempts have now closed it tight!

The detection of gravity waves, or gravitons, is going to be the Next Big Thing to happen in astrophysics. More equipment is coming online all the time – and excitement is now building for the deployment of LISA (Laser Interferometer Satellite Antenna), a joint ESA/NASA space venture scheduled for launch in 2011. This will use three spacecraft 5 million km apart in an equilateral triangle. The launch of a testbed spacecraft, LISA Pathfinder, is scheduled for 2010. This will carry sensors a mere 35 cm apart to confirm that the proposed technology will do the job. Excitement is now certainly building that this one will soon be cracked.

Who knows, it may soon be possible, next time your partner asks if Earth has moved for you, to confirm this with one of these teams!

CHAPTER 4
Medical Maladies

The original cache from which this book's material was drawn contained an entirely disproportionate number of problems in this category. Quite why is something of a mystery to me. Astronomical research can encompass a number of disciplines, but, astronautics apart, medicine is not one of them – unless you collide with something in the dark!

Yet our sympathies should be exercised for those who are afflicted by the various "complications" encountered here. Celestial observation can be a seriously stressful business, but we should try to avoid at all cost afflictions of the soul or body that might hinder potentially great astronomical discoveries.

Nevertheless, the history of science is peppered by stories of those who have fought physical impediments to win their spurs. Darwin, for instance, went on to make good use of his voyage on *HMS Beagle* despite continual seasickness, Galileo wrote his greatest masterpiece despite being blind at the time, and Isaac Newton fought frequent nervous breakdowns to do.. er.. something or other.

S. Ringwood, *Astronomers Anonymous*, DOI 10.1007/978-1-4419-5817-4_4,
© Springer Science+Business Media, LLC 2010

So, give thanks for the inherent strength by which you hold this book – and shed a sympathetic tear for those crushed by the following burdens.

(IT MAY BE OF INTEREST TO NON-OBSERVERS THAT ASTRONOMERS USUALLY CONFINE THEMSELVES TO USING THE SAME EYE AT THEIR TELESCOPES. THIS HAS REPERCUSSIONS.)

Dear Steve,

Doctors have told me that I am suffering from an untreatable condition called monocalitis. What is this? Is there a cure?

Seth
Charleston, South Carolina

Dear Seth,

This is caused by hours and hours of observation at the eyepiece of a telescope. The muscular effort involved in continuously straining an eye at the eyepiece leads to calcification of the eyebrow in the elevated position. This freezes the brow line at an angle of 45 degrees, making it look as if the eye is holding a monocle in place; hence the name of this invidious deformity. In serious cases it can prevent the eyelid closing during sleep. (This has led to the commonly held belief that astronomers always have an eye out for a clear night.)

I regret to say there is no cure, but options are open to you. The first is to actually wear a monocle. This fools people into thinking that this is the reason for the peculiar appearance of the eye. The alternative is to begin observing with the other eye. This will result in giving the face a more symmetrical appearance. Unfortunately, it leaves one looking permanently surprised. On the other hand, it could have the advantage of making the sufferer look constantly alert and not someone who can have the wool pulled over his eyes!

I cannot remember far back enough to say with confidence whether I ever had a problem using a single eye with telescope eyepieces. I am however fully aware that during public observing sessions, when those unused to using eyepieces have their heads thrust towards the telescope's portal, various difficulties become apparent. Closing the wrong eye is the least of it!

It is a curious thing that right from the start, once an astronomer starts looking through telescopes, he or she will favor one eye above the other. There appears no logical reason for this, since each eye can be equally accommodated by adjusting the telescope's focus. Without doubt, once the preferred eye has become used to being thus used, it becomes capable of discerning more detail than the other.

In the interests of science I once canvassed my local astronomical society on the subject and found that the vast majority used their right eye. I wonder, therefore, if this has to do with the different functional bias of the brain's hemispheres. As my wife has subsequently pointed out, I foolishly omitted to ask if their bias was associated with their left or right-handedness!

(IT SHOULD NOT BE FORGOTTEN THAT LASER EYE SURGERY IS STILL IN ITS INFANCY.)

Dear Steve,

I have read that faulty eyesight can be rectified by laser incisions made directly on the eye lens to change its shape (i.e., its focal length). Can this principle be taken to its logical (astronomical) conclusion? Is it possible to have your eyes operated on to give them a very long focal length, hence high magnification, without the need for a cumbersome telescope? It seems almost too good to be true that one could make planetary observations simply by opening the eyes in the right direction! I am planning to go for my laser operation soon. Shall I go for the telescopic option if it is possible?

Reg
Dayton, Indiana

Dear Reg,

It is not only possible, it has been done. My old friend Arthur Sprain took this opportunity when failing eyesight forced him to take remedial treatment. Since his eyes were to be modified in any case, he took the view that he might as well go the whole hog. Although the surgeon was at first unwilling to extend the focus to such an extent, he was persuaded to do so when Arthur said he would call his first naked-eye comet discovery after him. Enough said.

Needless to say, Arthur began making super discoveries nightly because his gaze was unhindered by artificial optics of any sort. The only disadvantage to him was that for normal everyday use he had to wear a pair of high dispersion goggles to allow him to see things close up (i.e., closer than 3 miles). Unfortunately he came to a sticky end when, on a day he forgot his goggles, he walked confidently off a high Atlantic cliff into the sea believing he was already striding across the opposing continent.

The remarkable thing about telescopes is that they are quite forgiving where eyeglass wearers are concerned. Those with long or short sight can dispense with eyeglasses entirely; additionally, many eyepieces are now designed to allow the observers to keep them on.

The factor determining whether eyeglasses can remain in place during observation is the distance between the eyepiece's exterior eye lens and the eye's best placement. This distance is called eye relief. Most eyepiece retailers will specify this value for the different designs they sell. Comfortable eyeglass retention usually requires a minimum of 15 mm.

Generally, eye relief distance decreases with smaller eyepiece focal lengths (i.e., greater magnifications). Therefore, eyeglass users may prefer increasing magnification via supplementary lenses such as the Barlow to swapping an eyepiece for that of a lower focal length. A generous eye relief also prevents greasy eyelashes from smearing body oils on to the precious optical surfaces.

Until recently, the adjustable focus of a telescope has permitted eyeglass removal only for those afflicted by long or short sight. But there is now an eyepiece accessory that allows correction for astigmatism, too. Now that is what I call far-sighted!

Of course, all this eyeglass nonsense can be dispensed with by the use of contact lenses. Those who wear them can happily ignore all the above. (Oops, maybe I should have said that at the beginning?)

(THE APPRECIATION OF FINE DETAIL IN A TELESCOPE'S IMAGE IS ACHIEVED ONLY AT THE ABSOLUTE POINT OF FOCUS. THERE ARE EVEN DEVICES THAT HELP THIS PROCESS. ALAS, FOR SOME, IT IS A FORLORN HOPE ONLY.)

Dear Steve,

I suffer from that dreadful malady focuser's palsy where, in attempting a sharp image, focus knob adjustment always overruns in both directions. I turn it firstly too far one way, then overcompensate by returning it too far the other. As you know, in its more severe form, the high frequency repetition of this to and fro focusing becomes the affliction colloquially known as focus twitch. In rare cases, like my own, the brain's attempts at compensating for this vibrato blurring results in the front surface of the eyeball itself popping in and out in time with the adjustments. The resulting headache and painfully bloodshot eye often has me reeling with agony and leaning lopsided to alleviate the high blood pressure in the afflicted eye. My wife has now come to describe my return from an observing session as the entry of the hunchback of Notre Dame; adding that she has no wish to be my Esmeralda! Help me!

Felipe
San Carlos, Mexico

Dear Felipe,

This physiological disorder of the nervous system, more properly known by its medical name of 'Optica Tremens Nervosa,' has been part and parcel of astronomy since the seventeenth century. Even Galileo was said to have been a sufferer, although the damaged nerve fibers in his case were those of the elbow (his 'scope being a push-pull focus) rather than those of the thumb and forefinger affected nowadays. His 'fiddler's elbow' was hushed up at the time.

As to a cure, medicine is still in the early stages here. Injections of cortisone have proved fruitful, although symptoms recur as soon as medication is lifted, as has to be the case since adverse reactions to the treatment are onerous. Transcendental meditation has had a limited vogue as a remedy – the absolute concentration being the remedial agent in this case. The disadvantage occurs as TM expertise reaches its peak, with tripod and man, both cross-legged, ascending uncontrollably into the night air!

Particularly for rich field telescopes of low focal ratio, the focal point is an infinitesimally precise position. As explanation, consider that the focus depth of an f10 reflector is a forgiving six times that of an f4. On such instruments particularly, it is desirable to have focusers with dual gearing that permit 1/10th motion at least. There are even motorized attachments that now permit motion finer than this.

[It is worth noting here that this item was written before the condition known as repetitive strain injury rose to notoriety. Perhaps, during a long night's observational focusing, one should remember to change hands occasionally!]

(AS EXPECTED, ASTRONOMICAL AILMENTS CAN BE EXOTIC.)

Dear Steve,

I hear there is a nasty illness called E.A.R.S. (Eyepiece Acquired Retinal Syndrome), which may be passed on by handling dirty eyepieces. What are the symptoms of EARS, and should I wear any kind of protection when using unclean eyepieces?

Tanya
Rouyn-Noranda, Quebec.

Dear Tanya,

Undoubtedly the first piece of advice is not to observe with strangers. You simply do not know where they have been. And wash your hands and face afterwards. Despite measures such as this, EARS is getting a stranglehold and it is essential to watch out for the symptoms in the observers around you. These are: excess rubbing of the eyes while colliding with tripods, repeatedly staring into a fierce torch in order 'to scratch the back of an itchy eyeball,' an inability to spell the word 'gfgkjdv,' exhibiting a strange walk that entails squatting every five paces to break wind, and finally, overproduction of saliva when handling other people's eyepieces.

This affliction is now rife and spreading throughout the astronomical community like wildfire. ... Damn it, why can't I spell 'yuisdbk'? I'll try again. 'Rdfyedsj.' Blast! 'Uxdfdub.' Damn it!

Sorry, got to rush and see the doctor. (This may take me some walking at five steps at a time.) There is as yet no vaccine, so please take care! And when observing, don't forget – *always* wear protection.

Surprisingly, not much is written about the health dangers of dirty telescope eyepieces. Indeed, this paragraph may constitute its entirety. Until fairly recently, it was surmised that Galileo himself went blind through eye infections acquired though multi-person use of his various telescopes. In fact, research discloses that he had suffered eye problems since his youth – possibly glaucoma. Latest opinions are that this was subsequently exacerbated by cataracts caused by rheumatoid disease in later life.

Fig 4.1 A blind Galileo dictating his last masterpiece to Viviani [Painting by Tito Lessi]

Incidentally, contrary to popular myth, Galileo only improved on early telescope designs and did not invent the telescope. Apart from UK's Thomas Harriott, he was probably the first to use it for the purpose of looking at the sky – not at the courtesan ladies disporting themselves on a distant Florentine balcony. Allegedly.

Remarkably, it is now known that Galileo used telescopes only terrestrially for several months, before eventually wondering if they might show something useful in the sky! Doh!

(RETAIL FEVER CAN FLARE UP IN ANY ENVIRONMENT. AND THERE IS NO LACK OF THOSE WHO RUSH TO TAKE ADVANTAGE.)

Dear Steve,

I have an affliction that I desperately need to cure. I suffer from the overwhelming compulsion to buy new eyepieces. Every time a company advertises a new focal length or a different design range my fingers start shaking and I just have to have them. I sell anything I can to feed my desire. I have so far sold my car, half the household furniture, all of my wife's jewelry, my wife, and various bits of my body that are fairly unessential. I have even sold my pet dog. I have heard that there is a discreet organization called Accessories Anonymous that helps people like me. Is it worth joining them? Or am I doomed to continue this mad collecting? I have got so crazy with it that to buy more eyepieces I have even sold all of my telescopes so I can't use them anyway! Help! Help! Help!

Tyler
Rutland, Vermont

Dear Tyler,

I have every sympathy with what appears to be a very common illness, but I still find this affliction hard to understand. My own eyepiece collection comprises of three oculars. One is a Huygenian from the aiming mechanism of a First World War field gun. The second is a half-sucked glacier mint I found down the back of a sofa. The third was 'liberated' from a Galilean telescope stored in a London science museum. I find that these are more than sufficient for all my needs, and they easily outclass modern eyepieces.

To be completely honest this entire computer-optimized optical design, minimum dispersion glass, fully corrected technology sales talk is a load of absolute twaddle.....

However, I believe my eyepieces may interest you. They may not be sophisticated, but they're good...really good. Do you still own your house? Oh, these eyepieces of mine. You only have to fe-e-e-l the sensuous silky touch of the shiny brass barrels, peer through the glistening eye lens to see the billowy comfort of a fathomless broad apparent field of view. How about your savings? Used them up yet? Just rolling my eyepieces on the palm of the hand brings an upwelling feeling of satisfaction and superiority. No modern eyepiece can equal them in quality and looks. Just send property deeds and cash and they can be yours. They're here besides me; I'm looking at them now. They could be beside you. Soon. Don't trouble yourself with Accessories Anonymous. They'll only try to stop you having these wonderful things. I'm putting them in a cosy parcel now, just to save time. They're snuggled in gentle bubble-wrap, like precious eggs in a downy nest. Waiting. As soon as you send me all you have left in the world this beautiful set of marvelous optics will be winging towards you. Just think of the ecstasy as you hold them yourself. Caress them, clean them gently with soft cloth and expensive unguents. Send the money. Send the money now.

In the old days (and here I am talking about the 1950s–1960s) there were very few manufacturers and retailers dealing in telescopic equipment. Many amateur astronomers made their own from war surplus – and a few even ground their own lenses. In a world now awash with fresh goodies in constant upgrade, there is always a temptation to buy the latest design of telescope or eyepiece, etc. Technological advances are indeed improving the abilities of even modest equipment. But be mindful that a new eyepiece will not automatically make you a better observer.

Fig 4.2 Accessories come in all shapes and sizes [Image by the author]

When buying an eyepiece there are two properties to bear in mind, the apparent and real fields of view. The apparent field is a property of the eyepiece design and comprises the disc of light that you see when you hold the eyepiece up to the eye. The real field is the physical piece of sky that you see when you use it on the telescope. The latter is therefore a function of the magnification (in other words, the ratio of the focal lengths of the eyepiece and the telescope). The real field is the apparent field divided by the magnification. Thus an eyepiece with a 50 degree apparent field at ×100 has a real field of 0.5 degrees – i.e., the disk of the full Moon.

(NOT UNCOMMON!)

Dear Steve,

I am deeply fascinated by the night sky. I would love to be able to watch the stars, observe the planets, sketch the Moon, and gaze in awe at aurora and meteor showers. But I have a problem that prevents me from doing so. I'm afraid of the dark.

Dane
Lancaster, New York

Dear Dane,

What an idiot! I fear you are doomed to frustration. However, you should realize that dread of the night (Nyctophobia) is a deeply seated primitive emotion that had its part to play in the survival of our forbears. Danger then lurked in every inky black pool of darkness.

These days, there is nothing about in the night that is not there during the bright security of daylight – apart, I suppose, from the odd hungry wolf or bear. Maybe a few bats. And you'd be fairly unlucky to encounter a burglar or satanic axe murderer.

There's also very little chance of tripping up on some unseen obstacle, and even if you do, most hospital emergency departments stay open 24 hours a day. Ghosts (or poltergeists), of course, are said to generally keep to themselves, unless disturbed. But then, the odd paranormal man-ifestation has rarely hurt anyone, physically. Try not worrying too much.

Fear of the dark is indeed a primeval instinct of self-preservation. Although it may be no accident that civilization began with the use of fire, it was not perhaps its heat to cook meat but its light to banish the dark that was the first use.

There is no shame in this fear for those starting out in astronomy. I can assure those concerned that it is soon overcome through enjoyment of the vast and wondrous spectacle that opens up before them through astronomy.

If such fear cannot be overcome there is a simple remedy. Become a solar observer. It involves generally warmer conditions and socially acceptable hours, too.

(THE UNIVERSE IS A BIG PLACE – WITH A LOT TO LOOK AT.)

Dear Steve,

I suffer from that complaint which, although widespread, is still talked about in hushed embarrassed tones. It is an horrendous debilitating psychotic disorder that prevents astronomy being enjoyed. Yes, I suffer 'observational constipation'.

I get my telescope out under a clear sky, put an eyepiece in the drawtube, and prepare myself for a view of wondrous splendour, and then it happens. I cannot decide what to look at first. My impetus is effectively blocked. Though every nerve in my body cries out to continue, the very thought of making a decision to look at one object prior to another freezes my intentions completely, and I am unable to move an inch. There I stand, stiff as a statue, the butt of jokes and little sympathy.

I understand that in common parlance this is called astroblock. I find the condition is becoming so bad that I now only have to consider beginning an evening's observation for my motion to cease abruptly – regardless of what I may otherwise be doing. This can get very messy, I can tell you. I have tried the usual things like throwing away all my star atlases and Messier catalogs. I have even deleted my telescope's autoguide database. Yet still, as soon as I think something like "...tonight I'll look at the Ring Nebula, and while I'm in that area, Epsilon Lyrae, which is nearby. ...aargh" it happens, I'm stuck. Please help me.

Noel
Parma, Ohio

Dear Noel,

It's damned obvious what's required. You need a mental laxative. Your failure to decide on the priority of observations lies in their apparent equivalence in your mind. You have yet to formulate a rule that would enable you to differentiate between equally enticing views.

Try deciding on a priority, perhaps, by looking first at the object with the lowest right ascension, or the highest declination. Which object's name comes first in the alphabet? Which object did you think of first? Which is brighter? Ask friends which object they would like to look at. But first make sure they are not having the same problem! And try not to have just two friends with you. These and other methods are easy ways to help you out of your mental gridlock. But beware, some people who have success-fully countered this affliction have succumbed to its opposite – compul-sive-repetitive observing, commonly known as galloping eyeballs. So, what are you going to do? Carry on as before, or take my advice? Make your decision!

As in life, the sky is your oyster – and that oyster is gargantuan. But help is at hand. Modern GO-TO telescopes harbor databases containing many thousands of objects – and these are even updatable through the Internet. By the merest press of a button, your telescope will point at the object of your desire (although your instrument's optical abilities will still govern its visibility!).

There are some who quite deliberately solve the above problem by doing something akin to celestial stamp collecting. You will know that the eighteenth-century French astronomer Charles Messier comprised a list of 103 celestial objects "not to be confused as comets." Their collective observation within just a single night comprises a curious challenge called the Messier Marathon, a task entirely without scientific merit or purpose. Yes, there's no doubt about it. Astronomers are weird.

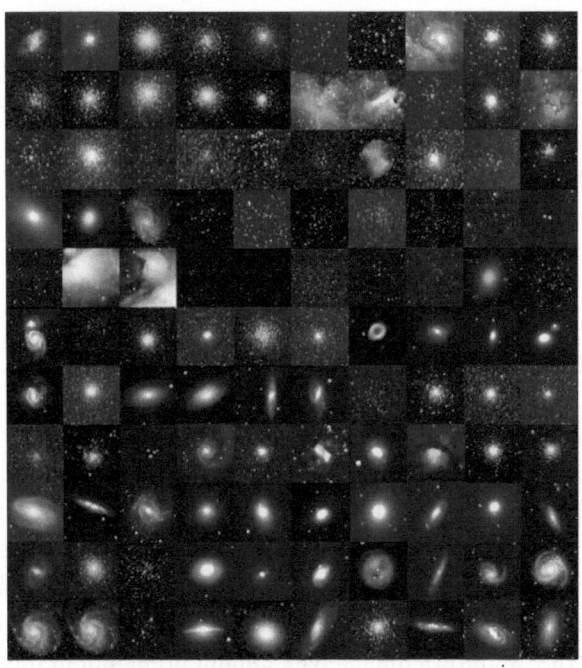

Fig 4.3 Collect the set – the Messier catalogue [Image: Courtesy of P. Gitto, NASA.images.org.]

The revised New General Catalog, a survey of galaxies, star clusters, and planetary nebulae, first published by J. L. E. Dreyer in 1887, now contains approximately 14,000 objects. Let's see you get all these in a single night!

(YOU MIGHT THINK THAT THOSE WHO STAND UP IN FRONT OF A GROUP OF ASTRONOMERS TO DELIVER A LECTURE HAVE THE HARDEST TASK. NOT ALWAYS SO.)

By email
From: John Darcy[jaydee@globular.com]
To: Dear Steve[doctorsteve@help.com]
Subject: Sleep disorder

I suffer from 'lecture sleeping sickness'. I acquired this ailment after being bitten on the leg by a lecturer for nodding off during his talk. I do not remember his name, but he did not like the color of our meeting room either. Strange chap. But ever since, I merely have to sit down to listen to a speaker's opening phrases when all of a sudden my leg begins to throb, and I fall instantly into a deep slumber. The other members of my weekly astro group now insist I sit inside a wardrobe in the corner of our lecture room to muffle my snores. Sometimes it takes a whole week before I wake up again – just in time to nod off once more. Do you have any recommendations?

From: doctorsteve@help.com
To: John Darcy[jaydee@globular.com]
Subject: Sleep disorder

Hi, John!

As a lecturer myself, I am often in the best possible position to observe others afflicted by this malady. I can mount the podium and simply take a breath to speak when, like dominoes in a hurricane, a sea of fresh faces turn suddenly into nodding hairy pates. But it is the lecturer who must take the initiative – sufferers themselves cannot break out of this sleepy spell unaided. I find that frequent loud mentions of MONEY and SEX can prevent the worst excesses of *audio somnolence,* as it is properly called. This can be difficult in a lecture on the event horizons of black holes, but I see this as a challenge. Otherwise, for the sufferer, it is simply a case of sitting at the back and enjoying a nice catnap!

The hardest task of any lecturer is to keep the audience awake. It's an uphill climb; consider the circumstances. The interaction is generally not a conversational one. The listener is not required to initiate or receive a response. A single voice can become monotonous. Sonorous. (Even snorous.) Worst of all, the lights are generally lowered or switched off – guaranteed to trigger sleep in an irresistible Pavlovian reaction.

Lecturers who want to be heard above the snores should be mindful to move around a great deal, make loud noises, and present exciting visual displays. It helps if the information being imparted is interesting in its own right, despite a boring delivery. It is also difficult to doze in the cold, so perhaps in the winter judicious use of open doors and windows will assist.

In support of the maligned audiences it should be mentioned that there have been isolated cases of the lecturers themselves nodding off mid-delivery. I cannot name them. I have spent their money.

(ASTRONOMY BOASTS SOME PRETTY IMPRESSIVE MALADIES.)

Dear Steve,

In writing notes at the telescope, I have found lately that I unconsciously launch into such romantically silly prose that afterwards I find it is complete gibberish. It used to happen only occasionally, but now I cannot complete a night's observation without being unwittingly forced into writing at least a thousand words of flowery drivel. The following is a recent example merely describing a single star, which I found in my notebook the next morning. "The central luminescence, sparkling shyly among the zealous guardians of the surrounding cluster, is a glowing pale ruby set among lustrous pearls. It draws the observer into its bewitching arms, like a lover's kiss, imparting a breath of heaven." *I was only looking at Aldebaren, for goodness sake! What on Earth's wrong with me?*

Keaton
Charlottesville, Virginia

Dear Keaton,

I fear that you have developed Smythe's disease. It is named after its first sufferer, Admiral Smythe, a UK astronomer who in the nineteenth century also suddenly found he had to write enthusiastically in lyrical terms of everything he saw. Unfortunately, in his time there was no cure. During his long life the affliction increased to such an extent that it infected not only his writing but his very speech, too! He was heard, on his dying breath, to relate that "With crystal eyes I have beheld this wondrous miraculous world and its exemplary experiences and now I forlornly fly to my fate in another." Whereupon the attending doctor decided he had heard enough and administered a lethal dose of leeches to see the old man off.

Happily there is nowadays a course of action. But I am afraid it is not pleasant. It consists of spending four weeks in the nearest library, inking out all the interesting adjectives from the English literature section. Fear not! Though drastic this deadly measure might seem to be, it crushingly relinquishes the tortured cerebral soul from its turmoil and lays to rest, in time, the expositions of meaningless moronity from which you ceaselessly suffer. Gadzooks! Methinks I shall verily have to do as like in the coming autumn season of the wilting bloom of life's travail.

The famous (nay, infamous) Victorian astronomer Admiral Smythe published his seminal work *Cycle of Celestial Objects* in 1860. His florid prose is well known and enjoyed by many. For instance, M11 was pronounced by him to appear like "a flight of wild ducks," thus giving it its common designation – the Wild Duck cluster.

In describing M19, Smythe says "It is of a creamy white tinge and is slightly lustrous in the center." He continues "The above nebula, and the whole vicinity, affords a grand conception of the grandeur and richness even of the heaven of heavens." Enough said. What a shame that modern texts are not written thus, for verily they would be wordy palaces of lofty celestial enlightenment.

But this underlines a real point. Although "an observation" can be a technical term dignifying the experience of simply looking at a celestial object and going "Wow!", an observation unrecorded is an observation wasted. There should always be a notebook by the telescope, if only to record which objects have been observed in that session. These jottings may seem insignificant at the time, but be assured they are a gold mine for the future.

(THE WHOLE FIELD OF EYEPIECE DESIGN IS RIVEN WITH PREJUDICE, PARTICULARLY AGAINST OLD-FASHIONED EIGHTEENTH-CENTURY EXAMPLES THAT STILL FIND EMPLOYMENT ON SOME 'SCOPES. THIS MAY BE WELL FOUNDED, AS SOME CONTAIN MILDLY RADIOACTIVE BARIUM GLASS. OTHERS HAVE BEEN AROUND LONG ENOUGH TO PICK UP OTHER DANGERS.)

Dear Steve,

Help me please! I am out of my mind with worry. Every time I handle my eyepieces I get itchy hands, and a dark mottled rash appears around my observing eye. What is it? The eyepieces are Victorian Ramsdens, dated 1833.

Micaela
Nashville, Tennessee

Dear Micaela,

Hardly surprising, is it? Ramsdens! What you have is an affliction called 'Panda eye'! Don't forget that, being almost two hundred years old, there has been the opportunity for perhaps thousands of diseased eyes to be pressed up against these oculars. During that time there have been some quite awful maladies lurking about. You really should confine yourself to modern, higher quality optics. Try saving up and buying sophisticated types such as Plossls or Erfles. You should find the allergic reactions cease once you get rid of the old lenses. This can be done fairly easily by enclosing them in a plain brown envelope, which is then dropped into a prepared dustbin at arm's length. Better still, wrap carefully and post them to me.

I have dealt with the cleanliness of eyepieces elsewhere, but would like to champion this most ancient (almost) of eyepieces. Designed in 1783 by Jesse Ramsden, these were an improvement on the mid-seventeenth century Huygenians (Christian Huygens) that preceded them. Both contain two plano-convex elements. In the Huygenian both the convex surfaces face away from the observer, whereas in the improvement they stare at each other. Although their apparent field of vision is small, Ramsdens have an advantage against modern eyepieces. There are only two pieces of glass. Light passing through the eyepiece is barely absorbed by the experience. In an endeavor that involves gathering as much light as possible this is a desirable property. Give them a try, particularly if you have a refractor.

Fig 4.4 A Ramsden eyepiece – the so-called "Comet" eyepiece of Broadhurst Clarkson, London [Image by the author]

(THERE ARE FEATS OF PHYSICAL ENDURANCE THAT THE BODY CANNOT ANSWER . . .)

Gee Steve!

I am an enthusiastic amateur astronomer and just love everything about space. And that is what has gotten me into trouble. The problem I have is ironically the result of a wonderful piece of luck that I had last month. NASA held a charitable auction, to raise funds for their 'Old Astronaut's Retirement Home' in Kentucky. This was to provide the residents with new leisure-time props, such as walking frame stick-on shuttle consoles, a replacement Apollo 13 DVD for the one that wore out, a lunar surface playground with bungee rope shoulder straps, and annual flights on the vomit comet zero-G plane. It cost me a lot of dollars, but I outbid a strange lady from the UK to win an old Mercury program spacesuit. OK, it had a stain down the inside of one leg, but I figured the $125,000 well spent.

When I got it home, it was like I was a kid again. I put it on for about three hours, walking slowly and 'talking to Mission Control' with all those shht sounds between each word. I hadn't noticed that it had already started taking over my life. At first, I was rushing back from the office, early, just to get out of my usual clothes and slip into the spacesuit. I'd leap about the apartment and bounce around on my bed pretending to spacewalk. It was great! Then I started wearing it to bed. Initially just the occasional night, you understand; but with increasing frequency. By last week I was thinking, 'What the hell, why waste a night?' So I didn't. It got to be every night, without fail. But to be honest, apart from the suit's complete lack of bathroom facilities (which made my nightly 4 a.m. visit a contortionist's nightmare) there was no inconvenience.

What's prompted my contact with you is that I have begun secretly wearing it to the office, under my business suit. I don't wear the helmet. That'd be a dead giveaway, right? Some of the staff have mentioned a strange crinkly noise when I move, but I just pass that off as a fresh pair of socks. But that means I can't use them as an excuse for the odd smell they've also noticed. I mean, how the hell do you dry clean a spacesuit? The other noise they have commented on is the strange cacophony of banging and cussing from my cubicle in the men's room each time I go in there. But I just can't bear not to be wearing it. The thought of not having the spacesuit on brings me out in a hot sweat – and that just makes it worse! It's been four days solid now, I'm getting hotter by the second, and even my eyes are stinging from the vapors rising through the neck ring. What the hell can I do?

Al
Danville, Kentucky

Dear Al,

Of course, the simplest answer would be to join the Mercury program, where your spacegoing garb would not be out of place, but you are about 50 years too late. I deeply suspect you would fail the selection process in any case. What we really need to do is urgently get it off, so to speak; otherwise you will most certainly expire from hyperthermia. There's also serious risk of liver infection due to the um...self-inflicted delay of relieving yourself. Aversion therapy is obviously out of the question – the devotion exhibited by the magnitude of your extravagance is obviously too high a hill to climb in the time we have. What we have left is replacement – dislodge your craving for wearing the spacesuit by inducing a less harmful, more powerful urge to wear something else. This something else will have to be light and permeable, so that your asphyxiated skin can breathe once more (even under your business suit) – and be so electrifyingly, overpoweringly attractive that you will be willing to rip the expensive spacesuit from your suffering body in a trice.

Tell me, what comes into your mind when you think of a string vest? Does it appeal? Just think of the air wafting past those little cells of exposed skin. The miniature eddies of musky aroma thus created. The slight sexual frisson of bondage inferred by the entangled cords. Think on it long and hard. But not too long and not too hard.

Finding obstacles in the way of getting to the bathroom, even those not of your own making, is nothing new in the history of humanity. Even the great intellectual endeavor we call scientific exploration is not immune. For what became perhaps the most famous bladder vent in history we need only revisit the space race in the early 1960s; May 5, 1961, to be precise. Astronaut Alan Shepard was sitting atop a Redstone booster in his *Freedom 7* Mercury capsule when, just 15 min before the launch of the US's first bid to put a man into space, there was a countdown hold that dragged out to an hour and 26 min. Unprepared for a lengthy stay (the flight itself was scheduled to last only 15 min!) nature began to call – then scream.

In a highly technical communication with fellow astronaut Gordon Cooper in the control room, Shepard announced, "Man, I got to pee." The trouble was that no provision for such a contingency had been made. No urine bag in the spacesuit! The only option would be to allow Shepard to simply relieve himself inside his suit. The medical team were at once alarmed that the unexpected fluid would short-circuit the bio-sensors inside the suit – with electrifying circumstances. The power to these sensors cut off, Cooper passed on the welcome news "Okay, Alan, Power's off. Go to it." The result was not as bad as you might think. The spacesuit he was wearing had a thick underlay that soaked up the waste quite quickly. He must have been looking forward to that shower afterwards, though! Thus there were two firsts that day. The first American in space, and the first (known) pee inside a spacesuit.

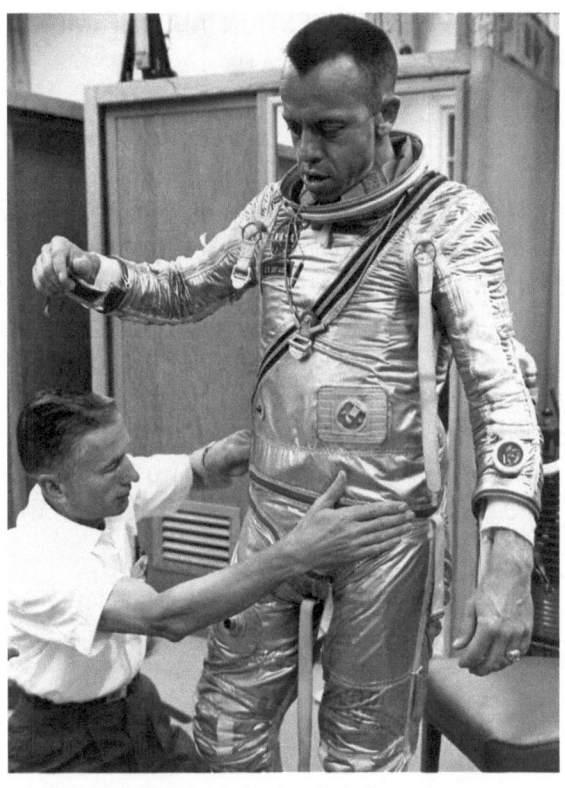

Fig **4.5** Alan Shepard prior to his *Freedom 7* flight, getting kitted out with everything except the one thing he really needed [Image courtesy of NASA.images.org. Marshall Space Flight Center Collection]

(BEING KNOWN AS AN AMATEUR ASTRONOMER AMONG YOUR SOCIAL CONTACTS EXPOSES YOU TO QUERIES NOT ONLY FROM THE EARLY COGNOSCENTE ENTHUSIAST, BUT TO THE INTERESTED LAYPERSON, TOO. THE LATTER CAN OFTEN BE MORE CHALLENGING, AS THE

CONCEPTUAL BASIS OF THE QUESTION MAY BE MORE <u>MAL</u>FORMED THAN <u>IN</u>FORMED.)

Dear Steve

I haven't stopped fretting since I saw it on the news last night. Of course, I was really enthusiastic when you astronomers announced that there would soon be a bright comet in the sky. Fantastic, I thought. A beautiful thing to enjoy in the evening sky. But that was before I found out that Earth was going to pass through its tail! What on Earth are we going to do?? I looked it up in an encyclopedia. The tail contains lots of horrible stuff, including cyanogen gas – cyanide! We shall all expire in our beds. Please tell me it isn't true. Will the end be quick or slow? I don't want to die, I'm only 89. Is there anything we can do to protect ourselves? Hide under our bedclothes or such-like? Quick, let me know, we don't have that much time.

Ethel
Dover, New Hampshire

Dear Ethel,

It's OK. Really. You will not expire from cyanide poisoning. So, gently, get out from under your bedclothes, please, and concentrate for a little while simply on breathing, slowly. The pollution of our atmosphere by these poisonous fumes will, after all, be the very least of our problems, as the very latest calculations of its trajectory have indicated. Impacts from the extended debris surrounding the comet's head will after all have already impacted Earth and started huge fires by the time the gas becomes an issue. (Do you live in a built-up area or near a forest?) And surely, you realize that the gravitational perturbations of the comet's close approach will in any case set off a string of strong earthquakes that will open up gargantuan cracks in Earth's crust so big they'll make the Grand Canyon look like one of the smaller wrinkles on your withered old face. (Thanks for the mug shot, by the way.) I'm pretty sure I can feel the ground shivering already. Can you? Even before that happens there'll be nothing left because the panic (looting, pillaging, and riots) will have destroyed all remnants of our civilization. The comet's destruction of Earth's surface will just clean it up a little. So, you see, you have absolutely no reason to worry about being poisoned. Nevertheless, I note that you are of...er...advancing years; certainly, your photo indicates imminent demise! Might I suggest therefore that, since I have been so helpful, I could have mention in your last will and testament? Just a brief mention, near the top. You know how it goes: "With thanks, I leave all my goods and chattels to Steve. . . " etc. Please don't wait, though. Do it soon.

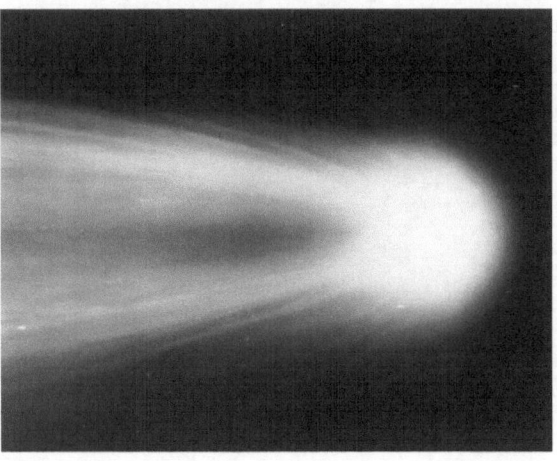

Fig 4.6 Halley's Comet, as it passed safely by in 1910 [Image courtesy NASA.Images.org.]

In 1910, Earth did indeed pass through the tail of a comet – none other than Halley's Comet, to be precise. The discovery of cyanogen in a comet's tail by Daniel Morehouse in 1908 was brought to the public's attention, unsurprisingly, by an enterprising and short-lived survival industry that sprang up during the approach of Halley's tail. Against a background of "end-of-the-world" hype, protective umbrellas, breathing masks, and headgear came into fashion. "Comet pills", too, were available that promised to counter the deadly effects of the cyanide that would rain down from the sky. Counter-pleas from actual astronomers that the intercepted gas would be so tenuous as to be almost non-existent fell on deaf ears. Of course.

We survived. Apparently.

CHAPTER 5
Guiding the Naïve

Neoteny. This is a term in developmental biology applied to any juvenile trait that is retained into adulthood. Consider the axolotl, an amphibian that retains its larval external gills into maturity; similarly, think of the relatively hairless skin of adult humans – unlike other apes, retained from infancy. And like the domestic dog, which retains its playful puppy stage throughout life, amateur astronomers are humans in a similar stage of behavioral arrest. They never lose the infantile wonder of the unknown that lesser mortals sadly lose as they get older. Astronomers are intelligently curious, because they are neotenous!

S. Ringwood, *Astronomers Anonymous*, DOI 10.1007/978-1-4419-5817-4_5,
© Springer Science+Business Media, LLC 2010

One day, quite without knowing why, larval astronomers will catch themselves looking wistfully at the night sky. The beauty of the celestial canopy becomes an awesome vista that will not allow the eyes to turn away. There wells up in their breast an unquenchable desire to know more, see more, care more.

It is at this fragile and critical point that thoughts turn inevitably to the purchase of a telescope. A mere cursory investigation of consumer opportunities then reveals a glistening array of tubes and tripods, each claiming to be the best window on the universe that money can buy. It is soon learned here that not only is the universe unimaginable, it can be unaffordable, too.

But also at this early stage, curiosity is outstripping sagacity at breakneck speed. Partial comprehension of this new intellectual landscape and the excess baggage of ill-judged conceptualizations mix in a swirling cocktail of feverish but uninformed enthusiasm. It is at this time that simple things poorly understood lead the unwary down very strange corridors indeed

Therefore, it is not a surprise to find that the newborn astronomical initiate is often desperate for information on what things are and how best to look at them. Those finding themselves in this very dark place cannot fail to notice a very bright light. It is a light that by its warm familiarity they do not fear. This light is the experienced amateur astronomer. Heretofore perceived by the novice as an anoraked weirdo, this well-traveled explorer of the cosmos becomes a font of useful knowledge. Sometimes, the information divulged is even correct.

(ALCOHOL HAS OFTEN PLAYED AN IMPORTANT PART IN THE HISTORY OF ASTRONOMY. MANY IMPORTANT TELESCOPES AND FAMOUS OBSERVERS HAVE BEEN FINANCED BY A BREWING CONNECTION, SUCH AS THE FAMOUS SEVENTEENTH CENTURY OBSERVER HEVELIUS, WHO FINANCED HIS ASTRONOMY VIA A THRIVING BREWERY. THIS TRADITION CONTINUES IN THE SHAPE OF THE CARLSBERG TRANSIT TELESCOPE ON LA PALMA. MOST ASTRONOMERS ARE KEEN TO

LEND THEIR SUPPORT TO THIS INSTRUMENT, OFTEN FUNDING IT ENTHUSIASTICALLY.)

Dear Steve,

Is it true that alcohol impedes your observational skills? I've been. . .er. . .overin-dulging lately.

Rory
Tampa, Florida

Dear Rory,
Yesh.

In small amounts it can debilitate your sense of color and detail, the ability to detect off-axis features in faint light sources, and cause problems in co-coordinating the brain's pointing accuracy with that of the telescope. However, I discover that in adopting a strategy of preparing for an evening's observation by consuming *large* volumes of alcohol (typically half a bottle of Scotch), you should find that one can not only discover things in the sky not yet seen by others, but also be the only observer ever to do so.

Make no mistake; alcohol does impede vision, even in tiny doses. While modest intake may make no difference to casual viewing through telescopes it will dent observations requiring higher acuity. Faint light perception, color sensitivity, and discrete resolution of the finest detail do indeed become impaired.

Contrary to popular opinion, alcohol will not "warm" the blood, either. The physiological effect of this poison is to open up the capillaries, thus speeding up heat dissipation at the skin's surface. Those planning a long winter's night observing are best advised to avoid it.

But, I hear you cry, did not dogged St. Bernards crisscross the Alps dishing out brandy to hypothermed victims, from a barrel slung from their trusty collar? If true, it is more likely, now that more is known of alcohol's properties, that a moment's inebriation was the last and only service the slavering canines provided.

Since the visually inhibiting qualities of alcohol are well known – and indeed have been since antiquity, it is surprising to learn that Galileo, no less, was a keen imbiber. A passionate gardener, it was logical that he should tend his own vineyard, and he drank its product with great enthusiasm. An accomplished connoisseur, he kept a good cellar to fuel frequent "wine-tasting" sessions with his friends. (His homeland of Tuscany, Italy, is Chianti country.) It is said that he described wine as being "sunlight captured by moisture," a phrase that speaks volumes of his obvious love for the substance. It was a love that caused some concern. His devoted daughter, Suor Marie Celeste (the "nunnified" Virginia Galilei), urged her father not to over-indulge – although her viewpoint as a nun living in the cloistered confines of the San Matteo convent somewhat detracts from her chiding authority. So devoted was he to his wine that on long journeys he always took some of his cellar with him – even when he was invited to Rome for "chats" with the Holy Roman Inquisition.

Following his condemnation in 1633, his consequent house arrest at his home in Arcetri, Florence, obviously took its toll on his cellar. By March of 1637, he had apparently drained his considerable stock (including at least 200 bottles of wine brewed during the previous 2 years!) and was sending out for more with some urgency. By his own admission, he states that he held back spending on all other pleasures in order to enjoy this one indulgence (although being in his late seventies one wonders what other pleasures he had in mind!).

(THE PHRASE 'CAVEAT EMPTOR' IS NO LESS SIGNIFICANT IN BUYING ASTRONOMICAL INSTRUMENTS THAN IN BUYING ANY OTHER TYPE OF GOODS. WHEN THE BUG BITES, THERE ARE PLENTY WHO WILL PROVIDE AN APPARENTLY SERVICEABLE GLASS WINDOW ON THE UNIVERSE THAT LATER TURNS OUT TO BE A MERE CHINK OF FRAGMENTED PLASTIC.)

Dear Steve,

I have been given a high quality 1" lens refractor for Christmas. It boasts scratch resistant plastic lenses, a finder (apparently somewhere in the box but I can't find it), an eyepiece providing a magnification of 2,500 times, a Sun filter specially commissioned by the Braille Society, a two legged tripod, and a really nifty white enamel tube. I have tried to read the instructions but am finding this difficult as they are in smudged pictorial script. Can you advise me?

Leo
Birmingham, Alabama

Dear Leo,

Heavens, it is vitally important that you take these steps before you go any further. Measure with utmost care the volume of space between the lenses (there's one each end of the tube). Put this space in a sealed plastic bag. Wrap the Sun filter in a handkerchief and smite with a 15 lb sledgehammer. Bury the glass fragments with due ceremony if you want to. Isolate the plastic lenses and place them in a pre-heated oven for 35 minutes. Although their subsequent removal from the oven will reveal an unsightly mess, the result will be far more useful than the original product. So far as the bipod is concerned, you will only find this useful for staunching a really bad nosebleed.

Take the enamel tube and cut longitudinally along the (obvious) seam. Flatten out using a rolling pin until all the wrinkles disappear. Mark out parallel dotted lines every 12 mm. Mark out parallel dotted lines perpendicular to these at intervals of 8 mm. Carefully fold along dotted lines, scoring with a blunt hacksaw blade beforehand if necessary. Check resulting cube of metal is 90 degrees true using a trustworthy setsquare. Place cube in the bin. Keep the bag of space. (It's the only bit really worth keeping.)

Although it has to be acknowledged that there are still examples of poor quality (nay, even scandalous) telescopes about, their availability is much less prevalent than previously. One of the reasons for this is that good quality mass-produced telescopes are now so affordable that cheaply made pretenders cannot compete! The golden rules are first to seek advice before purchase (from a local astronomical society) and second to buy from a recommended telescope supplier.

Suspect instruments, normally having an aperture of 50 mm or less, will have a single feature betraying their underperforming nature. The lens itself will have a delimiting diaphragm almost immediately behind it, preventing all but the central 10 mm or so to be employed. This is because only this portion of the poor quality lens is able to provide a (barely) usable image. This diaphragm should not be confused with field stops properly found in good telescopes; these will be further down the tube.

Galileo also used lens delimiters. This was for the valid reason that despite his best efforts the quality of his lenses did not permit their whole diameter to be used – and not because he was trying to pass off cheap 'scopes to a gullible public. In a typical example, although one of his lenses had a diameter of about 60 mm, less than 40 mm of the central region could provide a good enough image. The resultant increase in focal ratio did no harm either. His useful deployment of lens delimiters to sharpen images was a personal design innovation he "failed" to impart to contemporary telescope makers. Yes, it started that early!

(ASTRONOMY INVOLVES BEING QUITE INVENTIVE.)

Dear Steve,

I have terrible trouble keeping one eye shut while observing through a telescope. Doing what one should do (i.e., concentrate only on the view afforded by the eyepiece eye while leaving the other eye blindly open) leaves me with a hideous headache. If I simply attempt to keep the spare eye shut, my eyelid starts twitching. If I use my hand to cover it, my hand gets tired. What can I do?

Cristal
Burwell, Nebraska

Dear Cristal,

The radical and most efficient measure is of course the surgical removal of the 'spare' eye; but since this also impairs the use of your binoculars, this does have its drawbacks. It also means you don't have a spare in case of injury, etc. The methods you mention usually suffice for most, but in your case something else is obviously called for. One of the best ways of covering an eye is the wearing of a black balloon. This can be stretched over the head and perforated at the eyepiece position. Be sure to tuck your ears in properly because I have found that not doing so can afterwards seriously damage your hearing and make your ears stick out. Check that you can still breathe, too, as lack of air is detrimental to good observation.

In a similar vein, a tea cozy pulled down over the face, with the 'spout hole' conveniently placed, is also a good idea. A long lamented member of my local astronomical society was deservedly known as 'Lampshade Jean,' revealing yet another everyday item that can be pressed into similar service. An ordinary cloth bandana across one eye should be avoided at all costs; a friend of mine was recently observing the Moon in this way when an armed police unit suddenly leaped into his garden and dragged him away to jail as a suspected burglar!

Observing through one eye takes practice and can initially seem very awkward. Practiced observers are (as described above) able to keep both eyes open by zoning out the superfluous scene from the other eye. Keeping one eye closed is what most astronomers do to begin with, but this can become tiring during long observing sessions. "Pirate" eye patches are often used. And yes, the Lampshade Jean referred to above was a real person.

Of late, binocular eyepieces have become available that afford a restful and (some say) superior view. The disadvantage of these devices is that they eat up focal length (which can make focus unattainable), and their additional optical elements absorb hard-won light.

I am, in fact, rather surprised that the innovation of binocular viewers for telescopes has taken so long. Microscopists have had them for decades!

(SOME TELESCOPE MANUFACTURERS MAKE 'INTERESTING' CLAIMS FOR THEIR PRODUCTS. IT IS ALWAYS WISE TO SEEK ADVICE PRIOR TO PURCHASE.)

Dear Steve,

My daughter, now in high school, has expressed an interest in wanting a telescope for Christmas. The Lowdsacrap Catalog contains one, called the T40BS, which I believe would be ideal. But before making our purchase ($2,540) I thought it wise to check with you first. It is described as having a giant 40 mm lens, adequately protected by a 1.5 mm field stop and baffle. The tube is a shiny red to make it suitable for rapid dark-adaptation. The focusing rack is ultra smooth to prevent the plastic eyepiece from falling out. It also makes the point that it doesn't have a finder because there is no point in looking at an object with two telescopes when just the one is perfectly adequate. The Sun filter comes free with a list of internationally renowned eye surgeons. The tripod is boasted as having a special design that allows the telescope to bend with the wind to prevent damage to the fragile 'U' mounting. This all sounds well and good; but knowing the vast distances involved in the study of astronomy, my main concern with this instrument rests on whether the 5,000 times magnification will be sufficient to show anything useful.

Trouble is, tension in the house is climbing. My daughter wants to know she will get what she wants, I am dithering because I don't know what to get, and all the while Christmas is approaching like an express train. We're almost screaming at each other daily, and I've come out in a rash with the worry of it all. I've now got some tranquilizers, but good advice from you would be even better.

Tracy
Norman, Oklahoma

Dear Tracy,

Oh dear. I have to say that you would be wasting your money. Such an instrument would be less than useful by the single virtue that it is not expensive enough! Please consider the next model up, this being the T4OBSX. Although retailing at less than $4,000 ($3,999.99) its field stop is I.5 mm wider to enhance light grasp and also has a shiny blue tube instead. I heartily recommend it. If you are interested, I would be most happy to ensure that the executive director of the company sends you one C.O.D. without delay. I shall be seeing him at our company director's meeting tomorrow morning.

This note underlines the advice to always seek guidance prior to the purchase of a telescope. I have known of not a few instances where an unwary novice has acquired a first telescope that has completely quashed the generative zeal for astronomy. Those of us who are more experienced should always be on the alert for such instances so that this kind of disaster can be averted.

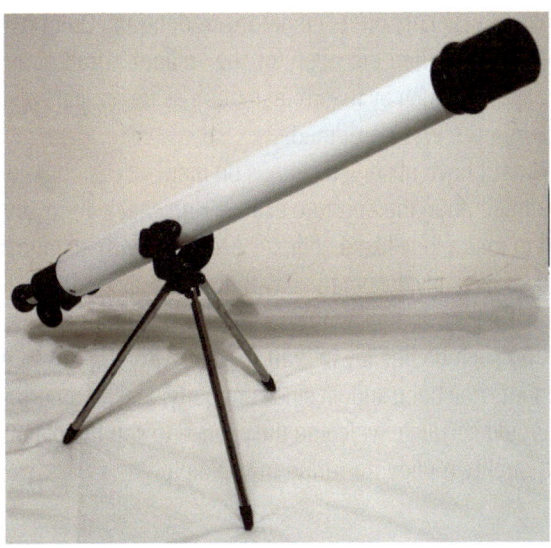

Fig **5.1** Promise unfounded [Image by the author]

The best advice that you can give anyone who is buying a telescope for the first time is to avoid like the plague telescopes whose chief selling point is high magnification. A lens can only deliver so much magnification, since as magnification rises the unchangeable amount of light passing from the lens or mirror has to work harder. A point is reached on any telescope where although (theoretically) an eyepiece is delivering a high mag, the telescope itself gives up and struggles merely to show a blurred entirely useless image. This limit of magnification is generally held to be about 50 times the aperture (in inches).

It is worth noting that Uranus discoverer William Herschel was reputed to habitually use magnifications in the thousands. The surviving lenses are little more than tiny beads of glass, thought to have been used only on his critical double-star work in order to improve visual separation of close pairs.

While we are talking of Herschel and observation in the same breath, let me say that I consider that astronomy's greatest observational blunder may have been Herschel's *non*-discovery of the planet Uranus. On March 13, 1781, he was observing from his garden in Bath, England, working on additions to his double-star catalog with a 6″ reflector at approximately 230× magnification. Near the star H Geminorum he came upon an object not recorded on his star map. Despite its fairly crisp appearance, he believed it was a comet! His paper describing that initial observation (and its interpretation) was published in the Philosophical Transactions of the Royal Society as an "Account of a Comet." It was only through subsequent observations over the following weeks made by Anders Lexell, Pierre Laplace, and others that suspicion was cast on its cometary credentials. Quite why Herschel gets the credit for his misidentified discovery of the "planet" Uranus therefore escapes me! Incredibly, it was as much by its orbital motion as its optical appearance that the object's planetary nature was confirmed.

I have to say that I have observed Uranus on many occasions – with instruments both larger and smaller than the aperture used in Herschel's discovery. At no time has it ever appeared to me to look like a comet – even one with an intense concentrated coma. It yields itself as a tiny uncometlike well-defined disk – albeit quite a tiny one. I can only assume that its initial non-reportage as a planet was a result of tacit assumption that no such bodies lay in wait for discovery. Do not forget, at the time, all the known planets had been known since antiquity, and no others were suspected. Nevertheless, I would certainly welcome the chance to see the planet through one of his 'scopes, or a quality replica, to refine my opinion.

(ENVY BETWEEN TELESCOPE OWNERS KNOWS NO BOUNDS.)

Dear Steve,

I am the proud owner of a 6" refractor. I have had it from new and taken extreme care of it – especially the optics. Understandably, I lose no opportunity to tell everyone at my local astro group what a great instrument it is. (The stuff they use is rubbish!) But a week ago I discovered, to my absolute horror, a tiny scratch on the outside surface of the lens. Of course, I at once sought the advice of the experienced members of the local society to which I belong. I was pleasantly surprised at their helpful response, despite the initial negativity I experienced when I first joined a month ago boasting about my wonderful telescope. When I phoned him the chairman at once confirmed the seriousness of the matter and counseled firm action. The scratch, he said, though minor, would scatter light throughout the whole instrument, rendering it useless.

The only way to solve the problem, he said, would be to grind the surface of the lens with heavy gauge emery paper until the scratch disappeared. I eagerly did as I was told. Amazingly it worked. I could no longer see the offending scratch. But this was because it was obscured by all the other, much stronger, grooves carved by the emery paper. When I rang him back, he countered my concern by saying that this was indeed what should happen and that it was just the first stage of the process. I was so relieved. The second step, he continued, could be described better by another of the society's members (an expert on this phase), and he gave me their phone number to call. I had trouble getting through to this number immediately because it was busy. But after a few minutes of trying I succeeded in getting through.

Strangely, this second guy already seemed to know what I was calling him about. He congratulated me on doing such a good job in the first procedure, before instructing me in the second. I was to choose a good quality power tool disk grinder. Setting this running in a vice, I was then to run the surface of my lens across it until the grooves from the first process disappeared. Again, it was remarkable how effectively and quickly this worked. The grooves had indeed disappeared. I then got back to the second member to find out how I now got rid of the deep gouges which had replaced them. Only the chairman, I learned, knew this next task.

Thanking the second adviser for his help, I then rang the chairman again. It was a terribly bad line. All I could hear at first was a sort of muffled barking sound, like an asthmatic wolf with hiccups. Dreadful! But eventually the line sort of cleared, although since the last time I talked to him the chairman appeared to have caught a bad cold. It was obviously upsetting him because occasionally he broke down and almost sobbed into the phone. Anyway, I got my next set of instructions. It transpired that the previous processes (without my realizing it) had in fact recreated the pristine new surface I desired. This surface was, however, now buried deep beneath the rough textural layer I had carefully formed. This rough 'skin' was now to be removed. This was to be done, I was told, by striking the edge of the lens a smart blow with a 5 lb sledge hammer, which I assumed would result in the coruscated layer flaking off in one piece to reveal a new pristine surface underneath. Apparently this had to be done with absolute accuracy and care, lest a faltering impact may do irreparable damage. The poor fellow then let out what sounded like an enormous sneeze before the line suddenly went dead. I felt loath to call him back, since he obviously needed some rest.

With advice from such knowledgeable people I had every confidence in carrying out this next task. Unfortunately, I must have gotten the impact angle just that little bit wrong. As soon as the hammer hit the lens the glass exploded into a thousand pieces.

Since I was not sure if this was a legitimate part of the process, I reluctantly called the chairman back. His wife answered, though she too seemed distracted by a dog being sick in the background. The dog seemed even sicker when I told her what had happened. She then had to hang up rather quickly – to call the vet perhaps.

So you see I am a long way into this restorative process for my lens without knowing what the next, apparently final, process is. All the glass has been carefully swept up and stands even now in a large jar in my garage. So please tell me. 'What do I do now?

Terry
Bristol, United Kingdom

By strange coincidence, I have already heard of your plight. Indeed, I have only just recovered. Suffice it to say that, unfortunately, you have not been best advised.

In the third process you should certainly not have used a 5 lb hammer. Flinging the damn lens from 17 feet onto a concrete floor would have done a much better job. Anyway. I suggest you melt the fragments in a cooking pot until good and fluid. Pour gently into a 6-inch pie tin (don't forget to grease it first). Once cool, push the resulting glass disk out and begin the process of grinding each face into the required curve. This formative process is described in my wonderful book Don't Follow My Advice – Find Out For Yourself. And if you end up with a lens with a scratch on it, please don't call me!

Suffice it to say that advice from fellow astronomers is generally of very high quality and altruistic – which is why it is always worth seeking. So far as optics are concerned, delicacy is always the watchword. Glass (or aluminized mirror) surfaces can be scratched by the gentlest wipe with a cloth inadvertently containing the tiniest speck of dirt. Delay cleaning optics until it is unavoidable. Habitual, frequent, and unnecessary cleaning will scratch optics. It's fate. The scratches will interfere with an instruments' light grasp and resolution.

The aluminum coating on mirrors will scratch far more easily than the surface of a lens. Surface dust, if you really cannot stand it any longer, should be removed with the softest brush available. And do not, whatever you do, breathe on the surface while you are doing it; damp dust is a great abrasive.

Lenses do permit a little more effort. Always brush off loose dust before doing ANYTHING, otherwise each particle of grit simply serves as an excavator. It can be useful to employ one of the various (compressed) air blowers available, but take care as some of these have been found to leave an unsightly deposit on glass.

If you use a liquid cleaner, do not pour or spray this directly on to the lens as the fluid may find its way into the optic's containment, with disastrous results. Apply it to a cleaning cloth sufficient only to make it damp. An easy way to test if a lens is clean is to breath gently on to it. The resulting temporary layer of mist on a clean lens will dispel immediately. On a dirty lens it will linger for a few moments, or even remain.

Lastly, avoid at all costs the irresistible temptation to finger stroke a camel's hair optical brush. It is programmed into our primate grooming behavior to behave thus, but the body oils that invisibly smother the surface of our skin will consequently be transferred to your precious lenses.

(SOME PROBLEM LETTERS EVEN HAVE SOMETHING TO DO WITH TELESCOPES!)

Dear Steve,

I have a problem aligning my finder 'scope parallel to the main instrument. No matter how much care I take, I can never get the cross hairs to intercept what I can see through the main eyepiece. Do you have any suggestions?

Daran
Sparta, Winconsin

Strange as it might seem, I do. I have never liked these new-fangled things! Finder scopes indeed! The seventeenth-century astronomer Hevelius always believed that sights alone were all that you needed to point a telescope in the right direction. If it was good enough for him, it's OK for me, too! I have updated this technique, though, to reflect today's more powerful telescopes and the rising trend to be one with your instrument. It is a technique I perfected during a residential astronomy course late one night – when luckily it cleared after the bar had closed. The idea is to entirely lift one leg over the telescope. Then, as you hug the tube with your arms, lay flat along it so that your body is parallel to the optics of the thing. With your chin hard against the skyward end of the tube, you then use your feet (if still on the ground) to shove the whole assembly towards the star of your choice – while of course maintaining a firm grip on the tube and your balance. Occasionally, this may impart some instability. Persevere. This method, called 'mounting the mounting,' is quite efficient. This is an isolated case of alcohol ingestion being a pre-requisite.

Of all the multifarious accessories that a telescope can be burdened with, a good finder is the most essential. A primary telescope has a field of view so small that sweeping the sky in the approximate direction of your target is doomed to failure. Always acquire the largest finder scope practicable, as this will ease the pursuit of fainter objects for your main telescope. On small telescopes, sights alone will indeed suffice, so Hevelius is still correct – at least up to a point.

Hevelius was a seventeenth-century astronomer living in Danzig (now Gdansk), Poland. Although his actual name was Johannes Hevel, Latinizing your name was a cool thing to do in those days. A rich brewer, he was the very last great pre-telescopic observer, but only just. Early telescopes were available, but in the task of mapping the heavens he decided simple sights were superior. It should be borne in mind that at this time the primitive telescope had only just entered science, while the proven technology of using sights and graduated scales had reached a pinnacle of perfection, especially in the expert hands of Hevelius. He produced a mammoth three-volume star catalog, MACHINA COELESTIS, between 1673 and 1679, using the various graduated sighting instruments erected on the roof of his brewery. The smell must have been awful!

Fig 5.2 Hevelius's observatory atop his brewery [Image from Machina Coelestis, 1673]

This observatory complex was destroyed by Danzig's great fire of September 1679. (Large conglomerations of wooden houses were often consumed by great fires in European cities in the late 1600s.) In addition to the observatory, undistributed copies of the final volume of his great work were destroyed – hence it is very rare. He attempted a rebuild but, a broken man and too old, the new observatory was never finished.

(BELIEVE ME, THIS HAPPENS!)

Dear Steve,

I have recently bought a 6" refracting telescope. My enthusiasm for astronomy has been dimmed, however, because it has not been performing at all well. I cannot see Jupiter very well. It looks just like a star, and its moons are quite invisible. When I look at the Moon, I can only see the larger craters and seas. Everything seems smaller than it should be. What is my problem?

Clark
Tampico, Mexico

Dear Clark,

Try looking through the other end!

There is a story that I do not believe to be apocryphal – having had it from the horse's mouth, so to speak. Many years ago, a teacher is purported to have purchased a reflecting telescope kit from a well-known (London) retailer. On trying out the newly assembled instrument, this customer bitterly complained that viewed objects were diminished, not magnified. It transpired that the primary mirror (that had a transparent base) had been fitted the wrong way around, thus presenting to the incoming light a convex diverging reflection, not a concave magnifying one. The teacher taught science.

(UNLIMITED ENTHUSIASM CAN BE WASTED.)

Dear Steve,

I have been an enthusiastic amateur for a couple of years and recently decided to build my own reflector telescope. I have already ground the 8" mirror, but on testing it, I discovered that the focus is inside the curve of the mirror. Can I still use it?

Ron
Isle of Man, United Kingdom

Dear Ron,

As a vase? Certainly!

There are still those who grind and polish their own telescope mirrors. The original drive for this practice was not only to enable the acquisition of a telescope when commercial manufacturers and retailers were few indeed, but also to save money on a telescope's most expensive component. It is fast becoming a lost art due to the ready availability of affordable instruments.

One of the earliest amateurs to build his own telescopes was, of course, William Herschel. In the 1770s, dissatisfied with the quality of available instruments, he turned part of his house into a work room in which he cast and ground his own mirrors. These were made with speculum metal, an alloy of copper and tin, with tiny impurities of zinc, arsenic, or antimony to whiten the result.

On one occasion, while attempting the casting of a particularly large mirror blank, the mould (made of horse dung!) sprang a leak – which turned into a flood. Upon hitting the stone floor, the molten metal caused the stone flagging to explode, sending flying shards of debris in all directions, sufficiently energetic to even strike the ceiling. The attending workmen, William, and his brother were all obliged to flee for their lives through the available exits.

Of course, we cannot pass by the subject of dodgy mirror construction without mentioning NASA's embarrassing error on the primary mirror for the Hubble Space Telescope. The intent of the space telescope was that the optics would be placed above Earth's blurring atmosphere to enable clear views of the cosmos hitherto only dreamed of by ground-based astronomers. But it was not

Fig 5.3 The most accurately inaccurate telescope in history [Image courtesy of NASA.images.org., Hubble Space Telescope Collection]

until the $1.5 billion telescope had been lofted into space in 1990 by the space shuttle *Discovery* that its 2.4 m primary mirror was found to suffer from spherical aberration.

The optical fault in the mirror had been caused by a faulty testing instrument, used to check the shape of the mirror's surface. Unfortunately this critical testing device, called a reflective null corrector, contained within it a component that was incorrectly placed along its optical axis by a single millimeter. This was sufficient to result in the creation of an imperfect figure on the mirror surface. Thus the HST possessed the most accurately inaccurate mirror in history. Ironically, the misplaced faith in the highly sophisticated test device allowed the introduction of spherical aberration, which would have been easily spotted by any amateur telescope maker armed merely with the edge of a razor blade.

It was not until a remedial visit by the shuttle Endeavor in 1993 that COSTAR (Corrective Optics Space Telescope Axial Replacement) was retrofitted and the HST began to fulfill its makers' hopes. Ironically a spare mirror had been made at the same time by another contractor. The spare was faultless. It was this one that stayed Earthbound!

(IF YOU TRY THIS AT HOME, I DISAVOW ALL KNOWLEDGE.)

Dear Steve,

I have just moved into a new house with a fantastic view of the southern sky. But a single large tree belonging to a neighbor of mine spoils this. It's rather tall and obscures a large fraction of my otherwise unimpeded view. I have tried broaching the suggestion that my neighbor might cut it down for me, but I do not think that he is very keen. What shall I do?

Pat
Chichester, United Kingdom

Dear Pat,

Colleagues of mine have come up against similar problems, and each has found their own answer. Praying has been successful in one instance I know of – the tree was hit by a bolt of lightning – but this cannot really be considered a viable or reliable option. Introducing a colony of termites to the immediate vicinity will effect speedy removal of any trouble in the area, but this tends to have adverse effects on other structures nearby (such as observatories). Five spoonfuls of scrubland poison administered to the root area will kill the tree in about three minutes flat, but as you have already discussed this subject with your neighbor it may look a bit suspicious if it should suddenly die.

There is another solution, previously presented to a packed and appreciative audience of the British Astronomical Association in May of 2008 (and the following July to the American Astronomical Society) by an astronomer who has been particularly blighted by this problem. This involved the use of metalized polyester film (mylar). The idea consists of making several parabolic reflecting mirrors of enormous dimensions, say 3 or 4 feet in diameter. This can be made simply by mounting the mylar on circular hoops, then allowing the breeze to impress a slight curve on the reflecting surface. These can be supported against the fence on your northern perimeter such that they reflect the Sun in its southern daily transit on to the unwanted tree. If sufficient care is made to ensure that the mirrors' foci intercept the arboreal barrier, the concentrated heat from the reflectors will slowly scythe through the tree's foliage as the Sun moves across the sky. The advantage of this method is that you are not responsible for the resulting damage, since the mirrors are unattended and you can hardly be responsible for the Sun's motion across the sky. I am sending you some mylar with my very best wishes. Let me know how it goes!

Failing this, may I suggest a method I have used to great effect at my home, where unfortunately I had the same problem. In this scheme I lay in wait until I knew that the tree's owner was away for the weekend. It only involved a little trespass. I cut the tree down, dug out the roots, planted a marigold in the same spot for good measure, and removed all evidence of the action. All this in the space of a single night! This left my neighbor somewhat confused on his return, as I stubbornly maintained there had never been a tree there in the first place.

By a cruel fate of celestial geometry, satisfactory observation for northern hemisphere observers of the Sun, Moon, planets, and many other objects of interest require a good view towards the southern horizon. This direction is easily discovered. It's where the neighbor has planted most of his or her trees.

Fig 5.4 Even big observatories can have trees in the way (Palomar) [Image courtesy of Greg Redfern (2003) JPL NASA.images.org.]

In a story that is retold ad infinitum, no sooner has the garden observatory been built than it seems fast growing trees are planted to blight its use. After years of ineffective complaining, the law in several countries is now catching up. It is now possible to have trees pruned or removed if it can be shown they are detrimental to the access of available light – or too close to a boundary. Recourse through the law is one thing, but a gentle approach direct to the neighbor concerned can save a great deal of money. Invite your neighbor over for an observing session – particularly when something is really worth looking at, such as an eclipse or close planetary apparition.

I should add that the last solution suggested above is based on a real event enacted by a known astronomer. Although the statute of limitations must be protecting him by now I will not name him, even under torture. Money could change my mind, though. . ...even better, money might stop me.

(STEADY THERE STEVE, WHOA..!)

By email
From: Lucy Hadron[lhayway@globular.com]
To: Dear Steve[doctorsteve@help.com]
Subject: Discovery

I am ten years old and I have discovered a new planet with my 30 mm Toyaquato telescope, please can you let everybody know. Also, can I name it after my mum? Her name is Agnes.

From: doctorsteve@help.com
To: Lucy Hadron[lhayway@globular.com]
Subject: Discovery

Hi Lucy!

It seems highly unlikely that you have found a new planet in the Solar System. This would only be possible with one of the giant professional telescopes and most certainly not with your miserable puny pathetic excuse for a Galilean optik tube, so there!

Amateurs do discover things, like new comets, stellar novae, and asteroids. As a group, amateur astronomers celebrate these achievements in the face of awesome professional competition. But it is an unfair battle – in favor of the amateurs!

Big observatories are pressed to observe only those objects for which their mammoth optics are needed. The field is therefore left wide open for the vast army of amateur enthusiasts who are constantly searching with far smaller telescopes the open skies left unguarded by their leviathan counterparts. Amateurs are even in the vanguard of lunar topography, through stellar occultation observations that reveal the shape of features at the Moon's limb.

One of the best known comet hunters is Bill Bradfield, an Australian who holds a record 18 discoveries to his name. Fourteen of these have been discovered with a mere 150 mm (6″) f5.5 refractor.

One imagines that discovering a new object can be one of the most wonderful things that can befall an amateur astronomer. Yet, in at least one instance I know of, it can have unexpected consequences. A UK astronomer of my acquaintance (I'll spare him further annoyance by keeping this anonymous – OK, he paid me) was innocently imaging a galaxy when there was subsequently discovered a small trailed image within the frame. The characteristics of those few recorded photons indicated the motion of a chunk of rock between Jupiter and Mars – a minor planet.

Of course, it was an event that admirably demonstrated that amateurs could do useful work and make a real discovery. Many hundreds of such minor bodies are spotted each year – and in that respect, the observation was unremarkable. It was quite rightly briefly promoted in amateur astronomy circles as an example of what could be done with moderate equipment. However, it came to the notice of the local press 2 weeks after its brief mention on local radio and from then on it snowballed. Over-eager "translation" of the object's diminutive status by creative reporting rapidly got out of hand. Before long, the national papers were announcing not the discovery of a small asteroid but a new major planet! Headlines were even accompanied by the rather patronizing and untruthful comment that at the time of discovery the astronomer's wife had simply offered him a cup of tea in lame response. (He doesn't drink tea.) Subsequently national TV beat a path to his door, for an interview which was subsequently given a certain editorial interpretation for maximum impact. Worldwide syndication of the story followed, with reports appearing as far away as Australia and New Zealand. To the utter bemusement and swiftly failing humor of the "discoverer," there even came a proposal to appear on an American TV talk show – with an offer to be flown across the Atlantic by Concorde in order to do so – an offer he declined.

At the center of this hype, our innocent astronomer became the target of unchivilrous ire from other amateurs who should have known better. I can imagine that pretty soon the pleasure and excitement of discovery was very much eclipsed by the problems that the overanimated media created.

One hopes that the irritation of those few short weeks are well behind him. But I can well imagine that the next time he finds a "planet," he might just keep the news to himself!

(BEFORE YOU EVEN BEGIN, CHOOSE THE RIGHT ENVIRONMENT.)

Dear Steve,

I have made the most awful mistake. When I bought a house recently, I was careful to ensure that the backyard was large enough for my planned suite of observatories. I was looking forward to moving in. This week, I did so. What a disaster! I have found to my horror that at night, the sky is drowned out by the arc-lights used by the commercial sports complex – just next door. It's a complete white-out! What can I do? Can I sue for invasion of darkness or something?

Milton
Chicago, Illinois

Dear Milton,

Mmmm. How is it possible for you to acquire enough money to purchase a home (apparently a large one) with earnings governed by such an obviously challenged intellect? Did it not occur to you that such a business in close proximity might have consequences? Such colossal vacuity can only be wondered at. Certainly, I have no sympathy – nor solution. Damaging their night lights with rifle shots can only be a temporary (and criminal) measure. And it'll be pretty obvious who the culprit is since damage to the illumination would be followed immediately by your very visible construction of astronomical observatories just the other side of their fence.

OK. Tempering my response just a little (I'm a nice cuddly guy, really) you might consider the obvious alternative. Concentrate your interest on solar observing instead. Other bright objects such as the Moon, Venus, and Jupiter are also accessible during the day and don't require a dark/night environment. I guess, stretching those strained gray cells even further, you could (ahem) consider moving again. But no nearer to me, please.

When my wife and I were looking for a home, almost 20 years ago, we were a source of great amusement to the local real estate agents; for we would stipulate that our first viewings must be made on fine days after dark. The local realtors, too, must have thought us odd — perhaps suspecting that incipient vampires were moving into the neighborhood. At each first viewing, we would eschew invited tours of the house (to the seller's puzzlement), but instead march immediately through the house into the backyard for studious examination of the local light pollution. If we were at all satisfied, we might then deign to examine the interior of the property. My wife blames this rather unbalanced assessment methodology for all the problems we have subsequently had with the house — and consequently squarely at my feet alone, as the ploy's sole perpetrator.

(A PROBLEM SHARED IS A PROBLEM DOUBLED.)

Dear Steve,

Putting it bluntly, please help save my marriage. I'm driving my wife crazy, but please tell her I'm doing something really important. Then maybe she will understand and let me get on with it. I learned from my astronomer friend last week that some mass has gone missing. I was aghast. I don't know much about astronomy, but this 'missing mass' problem now worries me a lot. So I thought I'd help. Don't quite know what this missing mass looks like, but I've been checking all the nooks and crannies in my home and garden — but so far I've had no luck. I understand it could be dark, so I'm particularly checking suspicious holes in the garden, the gaps behind the clothes in our closets, and those difficult to reach back corners in our kitchen cupboards. All sorts of places.

I had really high hopes for the cellar, but had no success there, either. I figure it moves about a lot, so there's obviously a need to keep rechecking old hiding places. There's been a few times when I thought I've found it, in the shadow behind a vase or picture frame. But when I lift the object away, both the shadow and the dark matter disappear. It's really sneaky, isn't it? My wife is not helping. She was amused at first, but as the days have passed her indulgence has inexplicably evaporated. You'd think she would want me to get involved, do something worthwhile for a change, but no. She doesn't understand that I can't stop now until I've found some. Every time I think I am getting close to missing matter, she screams and pushes me away from whichever cupboard I am emptying or throws herself in my way. Replacing the broken crockery is getting expensive! Could you please tell her, on my behalf, that this search is fundamentally essential to the understanding of the universe?

Tony
Cambridge, United Kingdom

Dear Tony,

Listen, let me try and break this to you gently. You have it all wrong. Your search for the missing mass is entirely misconstrued and significantly wide of the mark. Considering the same problem, I long ago observed that if you leave food lying around, it soon goes missing. This, I realized, was a big clue. So one evening I left a cake on the floor and hid behind a chair to see what would happen. Late into the night, I heard a strange menacing clicking sound. This got steadily louder, until it sounded like a waterfall of plastic cocktail sticks. What terrifying beast of anonymous consumption was approaching? Then, scuttling across the floor there came a grisly procession of cockroaches. Scrambling over each other in a final sprint they fell upon my cake in a bristly ball of legs and antennae. After only five minutes, the tangle of bloated cockroaches seemed to suddenly realize that the only material they now held in their jaws was themselves. They stopped and looked around, quivering. For one timeless horrible moment, I thought that one of them had spotted me behind my chair. I froze, breathless, as he unmistakably shuffled an inch towards me and slowly waved his antennae in my direction. Just as my aching lungs were about to burst, he seemed to give an insectoid shrug and rejoined the rest – who were busy casting about for non-existent crumbs. As one clacking body, they marched off the way they had come. There was not a trace of cake left. Even the marzipan was gone. This, I realized, is what is happening to unattended mass. It is being consumed by cockroaches, disappearing into their evil little chitinous bodies. Of course, as the human population increases, so will their amount of waste. All this will be eaten by cockroaches. So, slowly, as our numbers proliferate more of Earth's mass will go missing, inside cockroaches. Across the universe, parallel evolution will produce identical scenarios. Missing mass will be present everywhere!

I have tried to get this seminal discovery published in academic circles but have met with some unexpected reticence. With a rich sense of ironic justice, I indulge my revenge by surreptitious release of cockroaches into the most prestigious scientific libraries. Of course I have heard tell of theories involving exotic subatomic particles, but they have obviously not considered my cockroaches!

In his cheerful 1979 opus, "A Choice of Catastrophes", (publ. Simon & Schuster) Isaac Asimov looked far into the future of Earth's resource consumption – not by cockroaches, however, but by humans. He calculated that if the whole planet were given over to feeding humans – at the expense of all other living things – it could support a population of 1.2 trillion. At the current birth rate, he opined, this colossal number could easily be attained by the year 2280. Uncomfortably soon. But that ultimate head-count supposed "conventional" agricultural technology. So, being Isaac, he went further. What if we were to continue increasing our population unchecked (food presumably created via some matter conversion to edibility)? The current 35 year doubling of population would result in Earth itself being consumed in a mere 1,800 years, converting it entirely to nothing but a seething spherical mass of human flesh. The only way forward from that would be self-perpetuating Cannibalism! Yet, warming to his theme, he then speculated on an expansion into the cosmos. He found that at a conservative 2% per year increment of population, the gluttonous mass of humanity is capable of "devouring" the mass of the whole universe in a brief 5,000 years. In the face of such a threat, I hope the universe has a plan!

The problem of "missing mass" arises from an embarrassing disparity between the observed mass of galaxies and their gravitational behavior. Explicitly, a galaxy's suburbs orbit too fast – inferring the presence of undetected peripheral matter. The search for this material and its nature is being pursued by scientists with as much vigor as our troubled correspondent. Solutions range from large dark objects such as "failed star" brown dwarfs to mysterious phantom-like subatomic particles. One of the main contenders is the existence of the WIMP – standing for Weakly Interacting Massive Particle. (Physicists must take special acronym creation courses.) This group of particles are extremely heavy and numerous, yet very snobbish indeed in their dealings with common conventional matter – hence the problem in their detection. There is even speculation that the difficulty lies instead in an incomplete comprehension of gravitational theory, since the perceived disparity between implied and observed mass seems proportional to the size of the galaxy being observed. (The bigger the galaxy, the larger the percentage error.) Whichever way this turns out, we can be confident that it will not be mass that is actually missing, merely our understanding.

Fig 5.5 Apparently, there's more (mass) to a galaxy than meets the eye. A recent study of the Perseus Galactic Cluster (shown) by astronomers at the UK's University of Nottingham reveals dwarf galaxies embedded in a gravitationally protective cloud of dark matter [Image courtesy of Digitized Sky Survey, NASA.images.org.]

(I AM OFTEN ASTOUNDED – AND OF COURSE HEARTENED – BY THE SEEMINGLY INSATIABLE MARKET FOR AMATEUR TELESCOPES. SOME ENTHUSIASTS, THOUGH, WANT TO RUN BEFORE WALKING...)

By email
From: John Eager[eagle@globular.com]
To: Dear Steve[doctorsteve@help.com]
Subject: Danger warning!

Hi Steve!

Can you help me? Nobody is taking notice of me, and you are my last hope. I have just started out in astronomy, but three months ago I bought my first telescope – a 3" reflector. I got a great deal, as it was on special offer through my garden tools catalog. Over the last couple of months I have looked through it quite a number of times, so I therefore now consider myself a seasoned experienced observer – like yourself. That is how I know we're in great danger. But no one is listening! We need to act fast!

Last month, I spotted a really interesting reddish star that was quite bright. Every few days since, when the sky was clear, I have looked at it. It's been getting brighter. Of course, I have now come to realize that this means it is getting closer – coming towards us. Just lately, it's been looking less like a star and more like a red blob – it's glowing! It must already be heating up as it approaches our planet. It is exactly like that film where this thing comes from outer space and hits Earth – so I know it's true.

I have phoned the local police about it, but they don't seem to be taking me seriously. I have contacted several government departments and left messages for them to contact me, but they are obviously checking it out before getting back to me – although, it's been some time now. I have even tried to prepare my neighbors for the impending impact, but I'm finding it increasingly difficult to talk to them because they are either not at home or seem to be doing their running exercises in the other direction when I see them in the street. I don't know how long it's going to be before this thing hits, but I don't think we have long. Can you please use your contacts and alert those who need to know? I can even give you an accurate position for the object. By nine o'clock in the evening, it's just above the second chimney to the left of Macey's Cannery.

From: Dear Steve[doctorsteve@help.com]
To: John Eager[eagle@globular.com]
Subject: Danger warning!

I don't have to tell you that of course I was most alarmed when I read your email. Be assured that I have already called the authorities and exercised what little influence I have to remedy this situation. The fact that someone like you can be free to walk around, trying to talk to people is bad enough; but to actually have access to your own funds with which to purchase a serious scientific piece of equipment is simply beyond the pale. I am pretty sure that by the time you finish reading this email the police will have your house surrounded. Please give yourself up quietly. Do you have any sedatives in the house? I suspect you might. If you make no fuss, there will be no mess. There's a good lad. Keep cool – and live.

In the meantime, stop what you're doing and listen. Carefully. If you cannot yet hear the clicking sound of firearms being primed, you may still have a few moments left to read on. The 'red blob' you have 'discovered' is none other than the planet Mars, which I am happy to report is in no danger (yet) of colliding with Earth. We are just making a close approach to it – something that occurs naturally about every 26 months. I understand, though, that in order to allay your fears of personal peril (as a result of an impending planetary collision), a smallish room has already been reserved for you deep beneath Earth's surface. You shouldn't be disturbed, as the mine has been closed for quite some time now.

. . . .Sorry John, I was briefly distracted. I'm right now in touch with the officer in charge of taking you in. He wishes me to talk you out. He's putting me through to your land-line now. Is your telephone ringing? Hope so. Please go and answer it, now. And don't forget, walk very, very slowly. Especially as you pass the window.

The paranoid among us will no doubt be delighted to learn that there are indeed objects flying around the Solar System that can hit Earth – and frequently do. These range in size from a small house brick to that of a mountain or larger. Be comforted to know that every effort is being made to discover and track these threatening inter-planetary projectiles. NASA itself operates a search and identify program charged with finding all of the celestial objects that might get near enough to Earth to risk an actual impact. It is called, imaginatively, the Near Earth Object program.

The program's quarry is not the familiar rain of meteors, those brief rapiers of light that we often see dashing across the sky. Often misnamed, meteors are merely the dying scream of tiny particles no bigger than a grain of sand burning up on impact with Earth's atmosphere. The prey of the NEO program is their parental source, dirty ice ball comets and their less watery (but equally roguish) cousins, the asteroids. These objects, collectively, have a huge scientific interest, since they represent the construction debris from the origin of our Solar System. A few deserve special attention because their orbits run across that of Earth or have the potential to do so – resulting in the rather disastrous consequence of an embarrassingly coincidental location.

So real is this danger that each discovered threat is given a score (called the Torino number) that indicates its danger status. This delightful indicator consists of five color zones, running from white (no foreseeable threat) to red (certain collision), with a parallel numerical scale of potential risk running alongside this that goes from a 0 score of low probability to a 10, which means get the hell off-planet as soon as you can! There is also another risk rating that incorporates data from longer-term odds, called the Palermo scale. I don't know, but I'm wondering if its Siciian-flavored title is a nod towards an Earth-strike's resemblance to a Mafia "hit"!

Close run things are reported regularly, I guess to ensure continued funding for the program. (Me, a cynic? Surely not!) Browsing idly through the most recent near misses, I note that we avoided Armageddon by a mere 49,000 miles on March 18, 2009. How fortunate that we still have Bruce Willis among us for reassurance. Forward planning is essential. Possible deep impacts of known objects are predicted well in advance so that we can book early to avoid disappointment. I would, however, like to point out that most of the recent near collisions of new objects have been "noticed" only *after* they occurred!

Nevertheless, from the far back of the hall, I raise a nervous arm in question. Just what is it that we are supposed to do in the event of these expensive astronomers finding a red zone mark 10 Earth-seeking mountain bearing down upon us? "Ah!" they say. "Fortunately, the technology we were naughtily caught developing for use on Earth's surface can now be harnessed to blow these potential planet-killers to kingdom come." The science seems to get a little hazy at this point. Of course, this may simply be another excuse for continuing the funding of terrestrial weaponry. But as I said, me, a cynic?

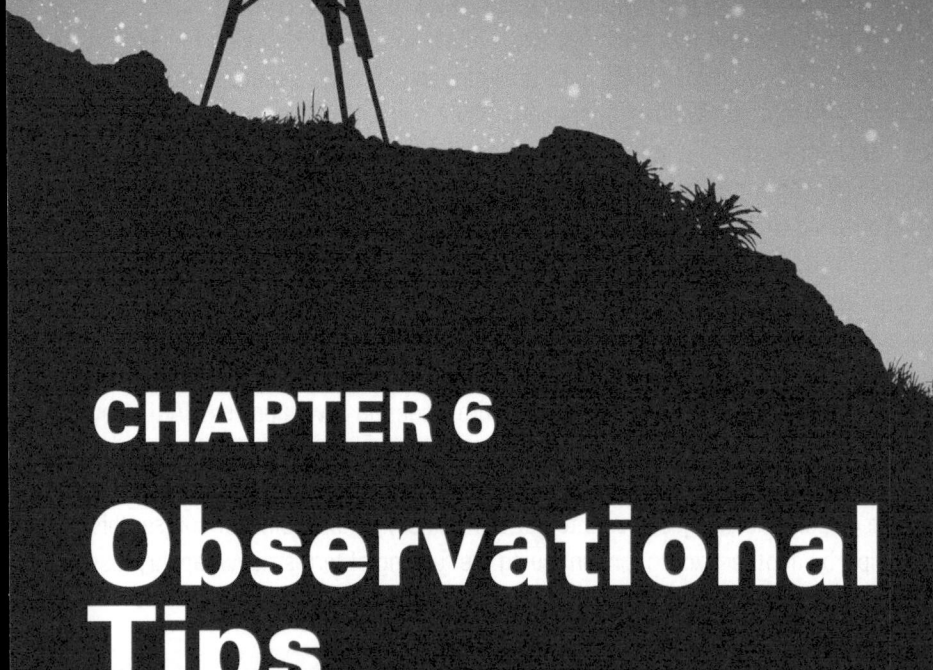

CHAPTER 6
Observational Tips

Generally, though not always, financial considerations dictate that an amateur's telescope is a good deal smaller than he or she would wish. Also, it may not be sited in the best place to have access to clear, dark, untroubled skies. Even those who disport themselves with powerful but mobile instruments often incur journeys to suitable sites that involve marathons of logistical arrangement.

The desire therefore to push at the boundaries, to gain every ounce of observational information from your instrument, is as keenly felt by the humblest amateur astronomer as the research professional. A vibrant two-pronged industry has therefore sprung up that claims to wring every useful photon from the amateur's optics.

The first is a set of physiological techniques that hone the fleshy end of the universe-observer relationship. This involves ploys like waiting for the eyes' slothful dark adaptation to complete in order to pick up the faintest trace of a distant and indifferent galaxy, or avoiding the ingestion of alcohol to preserve color perception. The

S. Ringwood, *Astronomers Anonymous*, DOI 10.1007/978-1-4419-5817-4_6,
© Springer Science+Business Media, LLC 2010

second is the commercial hardware of visual attenuators that strive to bludgeon the telescope into yielding up its maximum potential.

However extravagant and mystical these procedures become, they are all pursued with hopeful, often excessive, sometimes doomed, enthusiasm.

(DEVOTION TO MORE THAN ONE MASTER CAN HAVE UNDESIRABLE CONSEQUENCES.)

Dear Steve,

I find that going to work all day and observing during the night makes it difficult to stay awake for either. What can I do to stop my eyelids drooping at the eyepiece?

Myranda
Denver, Colorado

Dear Myranda,

Thankfully, telescope manufacturers have come to our aid. Many eyepieces now come with either integral or add-on eye guards. These short rubber cylinders are often used to keep extraneous light from entering the observer's eye. However, their true purpose is to assist those who literally cannot keep their eyes open. What is done is this. The observer puts his or her eye as close as possible to the eyepiece; then, taking the upper and lower eyelids by the thumb and forefingers of each hand, these are stretched forward over the eye guards so that the lids are clamped to the exterior of the rubber ring. This forces the eye to remain open despite the pressure of chronic fatigue.

A minor disadvantage to this measure is that the inside of the rubber eye guard can become fastened through suction to the eyeball itself. Should the head move suddenly while this is the case, the plucked eyeball may cause a bloody stain on the eyepiece optics – definitely a situation to avoid at all costs. The silver lining is that this can only occur twice.

Yet this does leave open the question of how you can still appear alert at the office the next morning. A long time ago, I went along to a waxwork museum and had them make a mask impression of my face with eyes that blaze with that well known "wide awake" look. Donning the mask, my hidden eyes can close, allowing me to sleep peacefully without anyone being the wiser! My sleepwalking is so good now that I wear the mask all day long. Nobody notices.

Many astronomers combat this problem by catnapping. Those who are office-bound during daylight hours may find this inconvenient. However, with practice, it is possible to spend part of the night observing and still get enough quality sleep to survive the day. It is not a common requirement though. I have (I believe) only three times observed throughout the night and gone to work the following day without a break for sleep. Apart from an initial grogginess, the body seems quite capable of resetting its diurnal clock in time for the following evening.

Current views on the minimum amount of nightly sleep required agree on approximately 8 h. British prime ministers such as Margaret Thatcher and Winston Churchill reputedly managed on 4 h sleep. Within scientific ranks, Thomas Edison could survive with just three. Albert Einstein blots the cerebral copybook somewhat by enjoying a sluggish 10 h every night. An interesting thesis might be built upon the sleeping habits of professional and amateur astronomers. Any takers?

(GREAT DAMAGE IS BEING DONE TO THE STELLAR VISIBILITY OF THE NIGHT SKY BY 'LIGHT POLLUTION.' PRESSURE FROM INTERESTED PARTIES HAS LED TO CONSTRUCTIVE ACTION AT LOCAL AND NATIONAL LEVELS. BUT FEELINGS CAN RUN VERY HIGH.)

Dear Steve,

I am writing on behalf of our group, the militant wing of the Black Light Bulb Brigade. Following recent adverse publicity by the imperialist press, we feel that the record should be set straight about our policies. We fervently believe that everyone has the inalienable right to see the stars at night. However, we certainly do not accept responsibility for the attack on the fluorescent tube factory last month. Several of our members were nowhere near there at the time. But we do actively promote and enforce the following, using reasonable means of persuasion.

1). *All artificial lighting, public or private, should be painted black. Any found without the required level of self-absorption will be removed.*
2). *Domestic light bulbs should be no brighter than 5 watts – any store found selling anything more powerful will get them smashed.*
3). *Any house found not using blackout curtains will be externally boarded up, doors included. Houses found using security lamps will have their addresses forwarded to local burglars for targeted challenges.*
4). *All torches should be licensed and must be covered by red cellophane at all times.*
5). *No one is allowed to light a cigarette in a public place between sunset and sunrise. Otherwise, for them, no sunrise.*
6). *Night-flying aircraft caught flashing navigation lights, or leaving an obscuring contrail in their wake, will have laser pointers flashed at them.*

We believe these are reasonable demands.

Anon
Johnstown, Pennsylvania

Dear Anon,

I regret to say that we are miles ahead of you! Having made little headway in recent years, the movement known as the Open Skies Campaign for Astronomical Regeneration (OSCAR) has decided to enter a phase of greater militancy to further its aims. We are currently working on the development of a low yield neutron bomb. After a brief flight on a powerful firework, it will be detonated low in the atmosphere, thus initiating an electromagnetic pulse (EMP). This, as you may know, will induce a power surge into every electrical circuit for miles around – and thereby switch off every damned light in the neighborhood. We're still wrestling with the knotty problem of ensuring the survival of astronomers and their equipment beneath the explosion. When we've done that, we intend to publish the bomb's design in the journals of the American Astronomical Society and the British Astronomical Association. Watch that space!

The Campaign for Dark Skies (based in Britain) is a worthy organization that has wrought miracles since its inception in 1993. It has been a driving force in both the deployment of more efficient (sky friendly) street lighting and government guidelines concerning night use of light. The UK's Institute of Lighting Engineers has issued a document "Guidance Notes for the Reduction of Light Pollution." I urge all readers to support the CfDS and similar organizations internationally in the fight to save our glorious night skies.

But does the general population care about their disappearing skies? Emphatically yes! When large parts of the North American eastern seaboard were blacked out by a power failure in August 2003, the full glory of the night sky was revealed across 9,300 miles^2 to 50 million Canadian and American citizens who had forgotten it was there. They marveled at the staggering and unexpected views of the starry heavens, including a totally forgotten Milky Way. Great protests and cries for more sensible lighting followed.

(LIGHT POLLUTION HAS A SOLUTION.)

Dear Steve,

I have a serious problem concerning my new neighbors. When I am observing at night I am hopelessly blinded by their recently installed backyard spotlight, which they insist on leaving on all through the evening and night. It floods the whole area and completely destroys any chance of seeing faint objects through my telescope and rules out any possibility of acquiring night vision. Is there anything I can do?

Shayne
Brampton, Ontario

Dear Shayne,

Legal proceedings are, I am afraid, out of the question, as this is not a 'nuisance' in that sense – although recent legislation raises hopes for the future. But I can assure you that once advised of the situation they will only be too pleased to curtail their use of the spotlight, especially if 'bribed' by a telescopic view of the Moon or Saturn (when the inconvenience of the light will become apparent to them).

If this has no effect, however, I would advise the purchase of an air rifle; a move which I personally have found to be most beneficial in my own surroundings. With a moderately practiced eye and a steady aim one can shoot out a light bulb at one hundred meters without the owner being able to know from whence the pellet came. But a really surefire method is to show your neighbors that, robbed of a celestial use for your instrument, your only recourse is to use the benefit of the illumination to observe THEM. In providing you with their various states of embarrassment visible only to your powerful telescope, observations will soon become available for the purposes of blackmail. Good luck!

As mentioned elsewhere, legislation is now being introduced in certain countries that concerns the domestic abuse of security lighting. Neighbors can illuminate their own environment, but it is now recognized that it is unreasonable for it to constitute a nuisance.

Rescue of the skies may come from another quarter. The cost of energy is rising fast. Oil is running out, "green" alternatives are barely able to provide a fraction of that required, and nuclear alternatives are unpopular. Therefore light itself is going to become a valuable commodity. Its waste (including scattered light making its way skywards and back again) will soon be seen as undesirable and severely curtailed one way or the other. So don't complain too much about energy prices, as they may actually save the night sky from oblivion.

A security light is the same thing as a security blanket; it provides an illusion of safety but little else.

(WITH THE WHOLE UNIVERSE AT OUR FINGERTIPS, HOW IS IT POSSIBLE FOR NON-DEVOTEES TO GET THE IDEA THAT THERE'S ANY TEMPTATION TO USE

OUR POWERFUL OPTICS TO OBSERVE CRUDE TERRESTRIAL SUBJECTS? I DON'T KNOW WHERE THEY GET IT FROM. I DON'T, I REALLY DON'T.)

Dear Steve,

When using my telescope on a frosty night, I always end up in a state of utter despair. This is because only minutes after I commence observation, the damn eyepiece mists up! How do I prevent this?

Chuck
Regina, Saskatchewan

Dear Chuck,

Caused by the condensation of warm air on a surface of cooler temperature, the answer to eyepiece misting is simple. Concentrate your energies away from your eyeball-heating neighbors' bathrooms and towards astronomical objects instead.

When this letter was first written, many years ago, there was no recourse once the lenses of your telescope had misted up on a cold night due to a layer of frosty

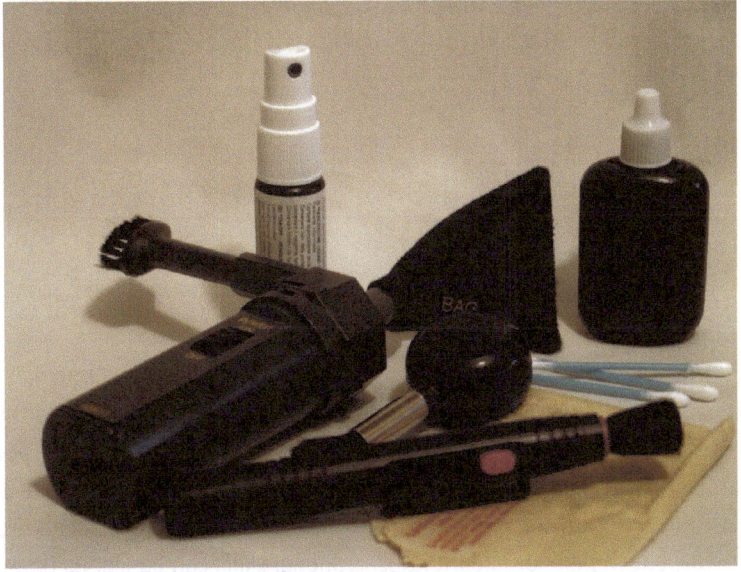

Fig **6.1** Typical lens cleaning kit [Image by the author]

condensation. Happily, it is now possible to surround your optics with heating strips. They maintain sufficient elevation above ambient temperature to keep the moisture at bay.

It also needs remembering that a clean, dust-free lens does not offer free-floating water molecules any catalysts upon which to condense. A dirty lens in a damp environment will accrue water as fast as any commercial dehumidifier. Therefore, when not in use keep the lens cap on the instrument. This ensures that you are still observing when all around you others are reaching for their sponges. It will also help if the telescope is brought to the air's ambient temperature BEFORE you remove any lens caps.

(THE ACCELERATION OF MODERN TECHNOLOGY OUTPACES EVEN THE MOST RESPECTED EXPERTS.)

Dear Steve,

I connected my telescope to a photon sensitive charge-coupled device incorporating super-cooled dense hydrogenous mercury peroxide within the atomic electron detector attached to the high-energy flux devamping ionizer. But I can't see anything with it!

Will
Atlantic City, New Jersey

Dear Will,

Pardon?

Sad to say, such is the advance of telescopic technology that old hands such as me can be left behind by the new terminology. I have but a single yet powerful defense. No matter how sophisticated your equipment, never forget that it is all about LOOKING, not the exercise of complex add-on telescopic technology. You can go that way, but you do not need to. Observational skills can be just as powerful.

Patience. Wait for the eye's iris to open to its fullest extent. This "dark adaptation" will enable your vision to become sensitive to the faintest telescopic image.

Unfortunately, owing to a cruel twist of physiological irony, this state is only achieved after more than 20 min of unadulterated darkness, yet is destroyed in the teeniest nanosecond by glancing at a misdirected torch beam.

As another example take averted vision. This is the technique of using the light sensitive "rod" cells at the retina's periphery, enabling observation of the very faintest telescopic images at the limits of visibility. Use of the retina's periphery is achieved by keeping attention on the desired object while deliberately looking away from it. This explains why astronomers go mad.

(THIS IS WHY I LIKE THE MORE ENVIRONMENTALLY COMFORTABLE SOLAR OBSERVING.)

Dear Steve,

Dedicated observer that I am, I frequently find that the onset of drowsiness cuts a planned long session at the telescope short. This is particularly trouble-some during the necessarily drawn out studies of eclipses and occultations. This must be a problem you have encountered yourself. Do you have any recommendations?

Elaina
Great Falls, Montana

Dear Elaina,

It helps if you get out of the mindset of only sleeping at night. I view any sedentary situation (day or night) as an opportunity to get some zzz's in. This can cause problems, as drivers who stop behind me at traffic lights can attest. But this approach offers so many windows of daytime oppor-tunity that I frequently do not sleep during the hours of darkness at all.

Stimulating drugs are, of course, another option. Caffeine, if taken in sufficient (i.e. prodigious) quantities, will not allow your body to sleep.

Sadly, it will not allow you to do anything at all, since a big dose induces manic restlessness, fidgeting, and uncoordinated body movements – in severe cases, feelings of persecution, real and imaginary.

Fig 6.2 Caffeine is a double-edged sword [Image by the author]

Oddly enough, the factor that most commonly curtails an observing session is not sleepiness, or even cloud. It's discomfort. This is mostly through bad posture or (when cold) inadequate thermal protection.

Get a good adjustable observing chair so that you can remain in situ at the telescope for long periods without strain. Do not be too proud to use a cushion. Once you are comfortable you will find that long observations of a single object will be effortless, not a strain.

Additionally, always overdress in winter. Cold penetrates more swiftly when you are static, so put on more layers than you would if you were just walking somewhere. That includes your feet! Raising your soles by one or two inches above the ground through wearing heavy insulated boots and thick socks will boost your winter observing time by hours. You can even invest in electrically heated waistcoats, gloves, and socks – available through most camping gear shops.

(RECORDING OBSERVATIONS WITH SKETCHES SHOULD BE THE AIM OF
EVERY OBSERVER. BUT WE ARE NOT ALL LEONARDOS.)

Dear Steve,

*I'm sorry. I just can't manage it. However much I try, I fail miserably when I try to
draw Jupiter at the eyepiece. If I illuminate my sketchpad with light it ruins my
dark adaptation for studying Jupiter. If I don't use a light, I see all the detail on the
planet but cannot see what I am drawing. Can you help?*

Kane
Idaho Falls, Idaho

Dear Kane,

I have a dependable remedy, one that should have been obvious to you.
For a start, stop trying to do two things at once. Can't be done. Shouldn't
be attempted. What you do is this. Most emphatically, do not use a light at
the eyepiece. Study the planet carefully and memorize the features. Then,
when you feel you have 'banked up' all the detail you can manage, rush to
your sketchpad in the lighted comfort of indoors and start drawing. Stop
as soon as you are unsure of what you are putting down on paper. Get
back outside to do some more observing. But it's cold, so it's important to
take a hearty swig of whisky before venturing outside again. Do more
studying. Repeat as necessary, not forgetting the essential whisky phase
of the operation. After several cycles of this, I can assure you that an
accurate depiction of the planet Jupiter is the last damn thing you will
care about. And you'll have a bloody good time t'boot.

Prior to photography, the recording of any telescopic image was by drawing at the
eyepiece. Galileo initiated this fashion when he started using his telescopes in late
1609. He sketched everything he saw – the Moon, Sun, the Pleiades, and the
satellites of Jupiter. Of course he was hampered by two things. Firstly, the optical
design he was using yielded incredibly small celestial fields of view – typically less
than 15 min of arc. Secondly, unlike his contemporary Rembrandt, he was no artist!
Yet his crude representations were seen as miraculous wonders when he released
them in his rushed but seminal work *Siderius Nuncius* in March of 1610.

By far the greatest astronomical controversy in history was created by a particular
series of drawings of Mars. During the close approach (opposition) of Mars in 1877,

the renowned Italian astronomer Giovanni Schiaparelli drew Mars in incredible detail, recording striated detail he described as canali (a term meaning "channels"). Fellow Italian observer Angelo Secchi appeared to confirm these observations. It was not long before the suggestion was made, most strenuously by the American astronomer Percival Lowell, that these channels were in fact artificial canals. Having built an observatory specifically for Martian observation, he began recording increasingly complex canal systems. This engendered further propositions that an advanced civilization was in the teeth of fighting planetary desertification by transporting desperately needed water from the Martian poles. It is worth noting here that in the late nineteenth century canals were the high technology of the period. Therefore, the stricken Martians were carrying out work as "advanced" as our own.

All this, despite strong doubts from such illustrious observers as Asaph Hall and Edward Barnard, who had the instrumentation at their disposal to detect evidence of canals or channels had it been there. But they saw none. Whatever the merits of the observations of Schiaparelli and Lowell, it did have a very telling affect on the general public's view of Mars and contemporary science fiction. The scenario of a dying Martian civilization was taken up by Edgar Rice Burroughs, H. G. Wells, Ray Bradbury, and Robert Heinlein. It is a theme that haunts science, real and fiction, still.

(TOTAL SOLAR ECLIPSES CAN BE LIFE-CHANGING EVENTS. BUT THEY CAN ALSO BE ABJECT DISAPPOINTMENTS.)

Dear Steve,

When I was young, the total solar eclipse of 1999 was a far off wonderful event that I could hardly wait for. Once I had decided that I was going to be under that lunar shadow at all costs, I built my whole life around it. In the thirty-odd years leading up to the event, I built up a small fortune to make sure I could have the very latest telescope equipment for its observation. I studied and gained degrees in physics and astronomy so that I could glean every ounce of new knowledge from the event. I bought a house on the Cornish peninsular of England, situated on the center of the eclipse's ground track to ensure I had an unassailable location.

I never married nor had children, to ensure that during the eclipse there would be no interruption. Then it happened, August 11, 1999. It rained. All day.

Where did I go wrong? I did everything. The targets, the preparation, the dedication. The medication. What could I have done to avoid this disaster? What do I do with my life now?

Charles
Penzance, United Kingdom

Dear Charles,

I cannot believe how anyone could have been so profoundly stupid! Astronomical pursuit in *any* location is determined by partings in the cloud, permitting brief glances at the sky above. Since most of the air received by the UK moves eastwards towards it from the very wet Atlantic, it suffers an unending supply of sodden air. (Note. I said sodden, not sodding – although it might as well be.)

I, too, was in Cornwall, using an umbrella instead of a telescope. The therapy afterwards helped. Indeed, I am sending you the address of my therapist. It may take a few months to get your first appointment, for I am only just getting over the disappointment myself.

Astronomers have traveled the world not only to escape the cloud of their native lands. They have also done so to make observations, from critical locations, that produce research results governing, for instance, the dimensions of the Solar System.

The relative scale of the Solar System was known in the eighteenth century and, with Kepler's laws, just a single absolute value was needed to deduce all the planetary distances. Enter the transits of Venus, those infrequent passages of this planet's disk across the face of the Sun. Theoretically, triangulation measurement of these events held the promise of such a value being determined.

Our hero, already glorying in the name of Guillaume Joseph Hyacinthe Jean-Baptiste Le Gentil de la Galaisière (Bill the Gentle, to his mates), was an eighteenth-century French aristocratic academician. He decided to observe the June 1761 Venus transit from the French possession of Pondicherry on the east coast of India. He sets sail in March 1760, but en route at Mauritius he finds that the French are now at war with the English. His captain refusing to proceed, Guillaume swaps ships and gets onto a French frigate heading for the Pondicherry coast. This ship is becalmed, and long delays ensue. When the ship reaches the Indian west coast at Malabar they receive word that those dastardly English have taken Pondicherry. Thwarted, the French ship heads back across the Indian Ocean to his back-up site at Mauritius – during which voyage our hapless astronomer sees the transit uselessly from a rolling ship!

Knowing that another transit (they occur in pairs) is due in 8 years, he stays in Mauritius. But once the war between the English and French concludes, he hurries back again to his observing area, deciding at first to observe the transit from Manila in the Philippines. For some reason he encounters what is termed "some hostility" from the Spanish authorities there. Rebuffed, he makes his way back to his original (1761) choice of Pondicherry. While waiting for the magic date, he builds a house, an observatory, and a school for the natives. He spends his time teaching the locals and studying the local fauna and flora. The transit day of June 3, 1769, dawns in brilliant sunshine, as has every day for the previous month. Then, out of a clear blue sky cloud approaches and completely blots out the Sun for the entire 2-h duration of the event.

He then contracts dysentery and remains bedridden for 9 months. Crushed by the wasted years of effort he then journeys home to France, during which time he is shipwrecked at least twice (versions vary). The sea having denied him, he is compelled to complete his tortuous return by trudging over the Pyrenees on foot. Upon his return to Paris he finds that without word of him for 11 years, he has been declared legally dead. His post at the French Academe has been given to someone else, his wife has remarried, and his property has been distributed to his heirs. (What a wonderful family reunion that must have been!) The lawsuit to recover it all bankrupts him. The king, taking pity on him, gives him back his seat on the Academe and he remarries – living for a further 21 years. He finishes his days almost destitute, living on the grounds of the Paris Observatory. There is a record of complaints against his (second) wife of her hanging out washing from the observatory buildings and making the place look distinctly untidy and unprofessional!

Those moaners returning home safely from a clouded-out eclipse trip should think themselves lucky and dwell on the fortunes of Guilluame Le Gentil.

I am truly astonished that this sorry tale of unremitting woe has not been snapped up by a movie mogul for Hollywood treatment. It has everything – seagoing adventures, battles, a race against time, courage and endurance, comedy, tragedy, a stolen inheritance, and an ultimate renewal with a love twist. I am happy to take my rightful percentage for the suggestion, with myself in the starring role as the disaster-prone anti-hero (of course!). I am standing by my phone right now, Mr. Spielberg.

Fig 6.3 Le Gentil's observing site at Pondicherry (Image: public domain). From http://www.transitofvenus.nl/images/pondicherry.jpg

(I HAVE OFTEN REGALED MY READERS WITH THE COMFORTING MESSAGE THAT THERE IS REALLY NO SUCH THING AS A FAILED OBSERVATION MISSION. APART, THAT IS, FROM THIS ONE...)

Hi Steve,

You have often taught your readers the cathartic properties of telling others of your misfortunes. I need to tell someone, anyone, about the unbelievable shocking events that recently befell me. You might have seen it on the news – but even so, read on. Please.

August 11, 1999, was to be not only the last solar eclipse of the twentieth century but the Second Millennium. It would be the astronomical PR exercise par excellence. Nations would grind to a halt. The world would be spell-bound. Grown men would weep. It was an eclipse that no respectable astronomer would miss. I sure did not intend to. I examined the eclipse shadow's ground track through Europe carefully. Reluctantly, as all the land-based observation sites looked prone to be cloudy during August, I decided that my only option was to view it from the sea.

How was I to know that August 11 would see the long arm of a low pressure weather front spread a thick finger of cloud along the entire length of the eclipse track? (Subsequent analysis now blames this on El Ni-moy. Nothing to do with the Pacific phenomenon of nearly the same name but apparently the mass evaporation of Star-Treky tears during an ill-timed TV showing of The Search for Spock!) Of course, the weather was the least of the problems faced by observers everywhere.

But I was having problems of my own. I had (sensibly, I thought) pooled resources with a group of other astronomers, and we had hired a boat and crew to take us out into the English Channel. We had every confidence that these seasoned seamen had the expertise to make it a successful trip. It went quite smoothly at first – that is, until we actually put to sea. We were not initially worried as we left the dockside, stern first. It wasn't until we continued in this fashion for three hours, in the wrong direction, that our suspicions were aroused. It was then that we discovered that our captain's maritime experience had hitherto been the collection of abandoned Spanish pedalos at a disused shoreside resort. This also explained his bizarrely paranoid insistence that the harbor pilot's tug remain attached to the ship even as we put to sea. We had to cut ourselves free!

Soon after, and not without incident (the ship's log will have recorded events I'd rather not make public in this letter), we were finally restored to the proper course. But it was shortly after our heated navigational discussion that the violent storm erupted overhead. By a cruel twist of fate, our shouting match had unfortunately drowned out the radio's storm-force gale warning that had apparently just succinctly said, "Get the hell outta there!" Honestly, Steve, if only we had turned back then.

However, keen to do our bit for observational science, we continued on our way towards the eclipse track, despite growing misgivings that anything would be seen beneath the strobe lightning and thunderous torrent of hail. Having almost certainly been robbed by the weather of seeing the total eclipse, we were still hopeful of using our photographic equipment on the gigantic waterspout that had formed just a mile to starboard. I can tell you it really was exciting to watch it turn from a twisted swan-like pillar of frothy water into a full-blown steely-gray colossus a mile wide. It was only when people began having to switch to their wide angle lenses that we realized it was getting closer to us – or, rather, us to it. Some of us also noticed that the base of the waterspout (which was now playing with a couple of air-suspended trawlers) had formed a swirling depression towards which our ship began to drift.

As has been related in the recent public enquiry, it was then that we had this great idea. After all, from the topside of the cloud deck above us, the moment of eclipse totality was fast approaching. Even beneath the raging clouds we noticed the growing darkness. Time was racing, as were our pulses. Although we astronomers were seasick, we were also desperate. We mutinied, locked the crew in the brig, and then executed this really great idea that one of us had thought of – not me, I hasten to add. We steamed the ship into the center of the maelstrom.

Yes, I know this seems rash now, but we figured that as a consequence the ship would rise rapidly up the column of water and appear, however briefly, above the cloud tops at its summit into an almost clear blue sky – when we might all too briefly catch a glimpse of the unfolding celestial event. At least, that was the theory (which, as you know, has subsequently been published in the recent edition of 'The Most Easily Avoided Disasters' by Noel Brown-Trousers). We decided to carefully time our impact with the waterspout so that our emergence above the cloud ceiling would be just on the onset of totality. (One of our astronomers was a professional meteorologist, so he helped in the calculations.) I can remember thinking at the time what a great idea it was. I still don't understand why. Maybe something to do with the session in the bar beforehand. Anyway, as you know, we set it up as well as we could.

In the radio room, an assistant was calling out the time signals over the ship's PA system. In the wheelhouse the smallest of our little group had been lashed to the controls and told to give the big red lever on the funny clock a hard kick to 'Full Ahead' when the radio room guy said so. On this pivotal signal, the ship would be thrown forward directly into the seething turmoil. It could hardly fail.

By now our marine tornado had become a howling heaving mile-wide monster of slowly rising water that dwarfed the ship. The liquid mountain rotated slowly before us as we inched towards it, reverse engines screaming to keep the ship from it until the right moment. The gnarled liquid trunk began to tower over us like a hideous menacing oak beneath a low black canopy. It was a fraught moment I can tell you.

Fig 6.4 Waterspouts are not good news for eager marine astronomers — unless you want to gain some height! [Image courtesy of Astronomy Picture of the Day Collection, NASA.images.org.]

Then, suddenly, heard faintly above the cacophony, there was the timing shout from the radio room. Before the echo had faded from the PA system, the ship's motors were thrown into forwards at full throttle. It lurched forward, throwing us all to the deck. The vessel, which we now know was far too heavy to rise up the column of water, broke into two halves as it hit the vortex.

Fortunately angular momentum threw our half of the ship away from the tornado, so we had time to deploy the lifeboat. Some of us then realized we were in the half containing the bar and had the public-spirited presence of mind to salvage some of its store. By the time we actually launched the lifeboat it was pretty full, what with 46 astronomers and the cases of Scotch and all. We did not even have room for the crew.

You would think that being drunk in charge of a lifeboat was a bad enough situation, but Steve our situation got worse. Despite pulling on the oars with all our might, the drag of the waterspout in mountainous waves was too powerful, and we began with increasing speed to be drawn toward the deafening roar of the spiraling cataract again.

Someone yelled a suggestion that we might sacrifice a virgin to placate the gods, but a swift roll call revealed that we did not have any on board. Then, the miracle happened. Just as we neared the base of the rising wall of water, the waterspout abruptly slowed, stalled, then slumped into the sea with such a powerful down draught we feared we might be sucked into the temporary depression that resulted. However, it almost immediately leveled out because of the rough sea. Yet by chance, the falling column of water created such a downdraft that astonishingly a hole appeared in the cloud cover directly above it. There above us, radiating glorious silver light through this celestial keyhole, was the Sun in the very last stages of totality! Incredibly, we had been having such a good time that we had not realized that only a minute and a half had passed since the ship broke up and totality was still seconds away from finishing. We were so stunned; you could have heard a wave lap!

I hurriedly recovered my senses enough to realize there might be time for a single photograph, just as the diamond ring effect was appearing. Despite the choppy water I set up my camera in record time and pressed the shutter. . . .just as the in-flying rescue helicopter eclipsed my view of the Sun. (And can I take this opportunity to once again refute the allegation that it was I who tossed the oar through its rotors and brought it down.)

I am still undergoing therapy for the shock of that moment. Part of the treatment is to write of my experience to you. Hope you don't mind.

Sandy
Hassocks, United Kingdom

Sandy, My Dear Fellow,

I am delighted that you have done so. I have myself been feeling low just lately, and your terrible tale is just what my doctor ordered. I believe your experience has a message to all astronomers who might consider taking part in eclipse expeditions. Unfortunately, no one has been able to work out yet what that message is!

Of course, actual experience can be just as vexing as the nightmarish scenes related above by our hapless seagoing astronomer. Indeed, this is the same mode of observation chosen by my family for the August 1999 eclipse. In the United Kingdom, the totality ground track clipped its southwestern peninsula. Good weather was indicated for the area – but not guaranteed. We decided to avoid the risk and attempt observation instead from a cross-channel ferry hired for the purpose by the publisher of the astronomy magazine to which I contributed at the time.

Our hearty ship, the P&O ferry *Pride of Portsmouth* left the UK's southern port of Portsmouth on the evening prior to the eclipse. On board with me were my wife, my 1-year-old son Martin, and our friend Dorothy. So, too, were several hundred of perhaps the scruffiest bunch of amateur astronomers I have ever set eyes on. Most carried airline cases whose contents probably contained a 35 mm SLR camera (1999 still not too far into the digital era!), a small catadioptric telescope, one of the 2,000 published guides to this eclipse, and . . . just maybe. . . a toothbrush. We, on the other hand. . . (those who have traveled with small babies will know the paraphernalia that has to travel with them). Add to this our tripods, cameras, and excessive clothing, and you may begin to suspect why my own small party resembled a 3-month expeditionary force bound for the Himalayas. It had taken two trolley-heaving trips at the quayside to get us and the luggage inside our first class cabin on the topmost deck. The magazine's other guests boarding either side of our cabin were mystic Uri Gellar and weatherman David Braine.

During the next morning we spent a short time (like everyone else on board) trawling the decks looking for a suitable observing site – like a flock of guillemots claiming nest space! So it was that minutes before the partial phase began, the four of us laid claim to a few square feet and began to set up our tripods and cameras. It was about that time when the fickle finger of fate began to wag in our direction.

Our friend Dorothy suddenly declared that her SLR camera had jammed. With the shutter frozen open and the film wind-on lever seized it was definitely in trouble. Fortunately a little inept fumbling on my part soon had the mechanism free, and Dorothy was thankfully soon ready to continue. Dorothy and my wife then left me briefly to return to the cabin to feed Martin. I continued with my preparation, but erecting the tripods was about as far as I got.

Aping the problem of Dorothy's camera, one of my own jammed, too, and I was forced to use a backup. I had just finished loading this with film when, quite suddenly, Dorothy breathlessly appeared. Apparently on the way back to the cabin, my wife had slipped on wet decking and fallen heavily while carrying Martin and both had, quite literally, hit the deck. Rushing to the first aid station I found them both very upset but relatively unharmed. I joked that since my son had almost bounced off the deck and into the sea, we should perhaps rename him "Barnes Wallis" after the famous WWII inventor of the wavetop bouncing bomb – a feeble attempt at restorative humor that sank without trace.

Fig 6.5 Preparations aboard the *Pride of Portsmouth* for the August 1999 eclipse in the English Channel. Wife Gillian and son Martin in the foreground, with our friend Dorothy standing by [Image by the author]

By the time we had recomposed ourselves sufficiently and returned to our observing site the eclipse was well underway. Dorothy had in the meantime repelled all encroaches from as yet unsettled astronomers. Back on deck, solar filters, lenses, cameras, and tripods were hastily assembled. After Gillian and I took our first snaps of the devouring dragon I continued trying to unjam my primary camera with increasing frustration and mechanical input.

All this time our ship had been slowly moving towards the central track of the eclipse and reached it just as the magnitude of the eclipse attained about 90%. At this point the captain cleverly changed course so that our path now matched that of the Moon's shadow. But it also changed the Sun's position from the deck. Oh how we laughed when astronomers who had spent hours setting up their equipment to ensure a clear line of sight towards the Sun now had to move en masse from the unexpected shadows cast by the bulkhead.

I stopped laughing very soon after. Although I still had a clear view of the Sun from my position, my line of sight now grazed the ship's funnel! Examination through my 600 mm telephoto confirmed my fear that the escaping heat was boiling the air so much that the Sun's image was a bubbling blur, never in the same focus for an instant. As every inch of the deck had by now been claimed there was nowhere else to go. Failure stared me in the face.

However, just minutes prior to totality, the captain trimmed his course a little and deployed the ship's stabilizers. It was just sufficient to steer my line of sight away from the disturbed air. The ship's stabilizers actually did a fantastic job, and in the calm sea my image of the Sun through the telephoto was rock steady! In the increasing gloom and chill, we waited. Thin patchy cloud had been slowly accumulating since first contact. It hung in the sky like a gathering of vultures, aware of the imminent death of the Sun. I wondered, not for the first time, if the fall in temperature beneath the penumbral shadow actually creates this sort of cloud if the humidity is high enough. But it would not be troublesome. We were going to experience totality!

Like the climactic scene from *Close Encounters of the Third Kind*, everyone on board was facing the eerie silvered light and casting weird frozen shadows. Incongruously for a ship at sea, it grew very still and quiet. Conversations were muted as if in church, hardly rising above the breeze that wafted them away. We were all worshipers with one purpose. All, as it were, in the same boat. The growing gloom revealed the distant lights of other ships around us, they, too, in tranquil reverence.

Quite suddenly the ingoing diamond ring was upon us, and a roar of approval rose from the ship. This reached a climax as the photosphere finally gave way and the corona emerged like a ghostly curtain around the Moon's pitch-black disk. A busy time ensued as (in a chorus of oohs and aahs) our cameras snapped away at the breathtaking spectacle.

Almost at once it seemed someone on the ship shouted it was ending. A second later the outgoing diamond ring was upon us, greeted by whoops and applause from the whole ship. In only a few seconds, like a hasty sunrise, it was broad daylight again. It was over – though not quite.

It was almost a week before I had a chance to unload the film in the cameras. On Gillian's camera there was a problem. The rewind knob rotated freely, indicating that all the film must have been wound on to the take-up spool and then been pulled from the film canister. This had frequently happened to me, and the solution was simple. Placing the camera deep into a knotted trouser leg (to make a Gerry-rigged dark bag) I opened up the camera back by feel.

It was only then that the full horror of the problem was revealed to me. I had been about to load Gillian's camera when Dorothy rushed to me with news of her fall with our baby. By the time we had returned to photograph the eclipse, I must have forgotten I had yet to load it. Throughout the eclipse, as Gillian snapped away enthusiastically, there had never been any film in her camera.

Gillian may forgive me. One day. After all, at the time of writing I am still alive. (Should anything suspicious happen to me . . .) All being well, I should have the best part of the next 25 years to make amends.

(SOMETIMES, SEEING IS NOT BELIEVING...)

Dear Steve,

I can understand that two telescopes of similar size might pick out different detail in a target – given slight disparities in optical design or condition. But can you tell me, please, if it is possible for observers to see different things through the SAME telescope? Something's happened that has caused me to doubt my observational abilities.

I was using my brand-new expensive telescope to show my neighbor and his friend the Moon the other night. I could tell they weren't very interested, but after I had pointed out the crater Tycho to them, my neighbor remarked to me that its central peak was turning blue. I told him not to be silly and checked for myself that this was not so. He looked again and repeated his claim. His friend, who oddly had not remarked on it before, confirmed the color although said it looked more aquamarine to him. I looked again and saw nothing – although it was difficult to concentrate as the other two kept making funny noises. I argued with them that I was an experienced observer and that Tycho's peak was definitely not blue. They kept looking and disagreeing with me. Then one of the guys finished looking, and to my short-lived relief announced that it wasn't blue. It was now green. The other pushed in excitedly and confirmed that, yes, it was now green.

I cleaned the eyepiece, I changed the magnification, and I checked the focus. I could see nothing, and they still said it was green. I suggested that maybe their vision was being affected because they were clearly unwell. They looked terribly flushed in the moonlight and kept bending at the stomach wheezing dreadfully, as though they had eaten something awful. Quite suddenly they said they had to leave, and they hurriedly returned to my neighbor's house. They must've only just made it because just as his door closed I heard the most awful noise – terrible sickness or something. Doesn't excuse the rudeness of their sudden departure, though. I checked out Tycho again and it still wasn't blue, green, or any other color. I'm now wondering how much I am missing. If two inexperienced observers are seeing stuff that I am not, what point was there in getting a new telescope in the first place?

Lloyd
Macon, Georgia

Dear Lloyd,

Let me hearten you by stating my well known mantra. Facts and truth are not a matter of democracy. The majority of a group claiming something is white, when it is actually black, does not make it white. The group may merely have a vested interest in calling a black object white to suit purposes that have nothing to do with scientific observational accuracy. Vague, unexpected hues on the Moon have indeed been recorded. But I don't think that applies in your case.

I certainly believe it was your fellow observers who were green, not Tycho's peak. I think it is fair to say they were pulling your leg. You could easily have verified this at the time by switching between blue-pass and blue-blocking filters – such as blue and red. Any coloration would then have been made very obvious – as would its absence. Nevertheless, might I suggest a little turning of the tables. Tell your neighbor that finally you did indeed see Tycho's coloration yourself – and proudly reported the observation to the U. S. Naval Observatory in their name as principle discoverers. Officials, you could say, are even now on their way to interview them. Stand back and watch them sweat.

Since the invention of the telescope, vague localized obscurations and pale colorations have reportedly been seen on the Moon. These sightings were scoffed at, surmised as being the result of atmospheric effects, poor equipment, or mistaken observations. The Moon, after all, was a dead, inactive body. This attitude changed on the night of November 3, 1958. Using a 50″ Cassegrain reflector at the Crimean Astrophysical Observatory, Dr. Nikolai Kozyrev quite by chance noticed something odd about the central peak of the crater Alphonsus. The usually sharp feature appeared to be obscured by a bright, hazy cloud. Unprepared for the event, he nevertheless scrambled a spectroscope to the telescope and obtained a spectrum of the strange mist before it dissipated.

There followed determined observational programs, mainly carried out by amateur astronomers, to keep a detailed log of such occurrences. Now called transient lunar phenomena (TLPs), brief puffs of what appear to be gaseous emissions from a strained lunar crust are now routinely recorded – particularly during the Moon's closer approaches to Earth in its elliptical orbit. Confirmed by seismometers left by the Apollo missions, the Moon may be dead, but it creaks occasionally – enough to vent a little senile wind towards the stars.

CHAPTER 7
The Expert

Although astronomers, amateur and professional alike, have long lifted their querying eyes to the sky, the bursting of the Space Age on the scene in the 1960s brought forth many questions from the curious populace. (Curious here meaning their interrogations, not their behavior, of course.) As able ambassadors of space science generally, astronomers therefore found themselves the interpreters to the lay pubic of all things non-terrestrial. So, together, we have explored the remote vastness of the Solar System and commentated upon the musings of philosophizing cosmologists. Consequently, astronomers

S. Ringwood, *Astronomers Anonymous*, DOI 10.1007/978-1-4419-5817-4_7,
© Springer Science+Business Media, LLC 2010

have been able to contaminate the less informed with unbounded and infectious enthusiasm for the subject (hopefully!).

The public's consequent perception of astronomers as authoritative but peculiar fonts of knowledge of all things astronomical (and beyond) engenders some very off-the-wall inquiries. On the one hand this is good. Debate and inquisition on all things scientific shows evidence of a healthy, intellectualizing, outward-looking society; on the other hand, ill-perceived conceptualizations and plain wrong thinking is laid bare for all to see. The Dear Steve letters in the following pages meet these curving balls with an unwaveringly straight bat, sometimes kindly, sometimes not.

(TV ASTRONOMY PROGRAMS HAVE BEEN TRANSMITTED SINCE THE 1950S. CONSEQUENTLY, A REGION OF SPACE MORE THAN 100 LIGHT YEARS ACROSS HAS RECEIVED THEIR RADIO-BORNE MESSAGES. IT SHOULD NOT BE A SURPRISE, THEREFORE, THAT THERE ARE ENQUIRIES FROM MUCH FURTHER AFIELD THAN EXPECTED.)

B453-98
Hagem fud

Giffdug muttn noruide deblebopp e hyplod. Artu cuvzags cundza., qi dehi jenrun-duaj massywust e jeawagsee dasd minkl ~Frowc. SCkisid dukcmuec d asd blaaaaa, tif e blaaaaa! V feShyns we fdfs?

Sht, nhjgf
sgfhsghg.

Dear Hagum,

Blaaaaa? NuBiusc fej muffs a dasd wee sagoff iff, fuyer mitt e logblatt!

New technology is discovering extrasolar planetary systems all the time, at an accelerating rate. Some of these have been found to be close by, relatively speaking.

Considering the content of planetary transmissions since the invention of radio, I am confident that their inhabitants will leave us well alone!

Yet the spatial volume of radio wave exposure since Guglielmo Marconi's first transmitted squeaks of the late 1890s is really enormous. Consider that the distance a photon will travel in 1 year traveling at 186,282 miles (300,000 km) per second is about 6 million million miles. (This distance is described as a light year [ly] to save wear on the "m"s and "l"s on typewriters!) Radio waves from Earth, streaking out into space in all directions, put us at the center of a terrestrial radio-filled bubble more than 200 light years across. So, what are the chances of alien listeners within a space of over 4 million cubic light years? Pretty high, I'd say. But would we understand their call?

(THERE ARE A FEW ASTRONOMERS WHO SEEK FAME AND FORTUNE AS POPULARIZERS OF ASTRONOMY. BUT THERE'S ONLY SO MUCH MONEY GOING AROUND.)

Dear Steve,

My friends say I have a knack for putting complex matters across simply. They also say that I have an immense knowledge of astronomy that I should be putting to better use. Moreover, since I have natural charm, wit, good looks, and no ego, I am seriously considering a dominant position in TV astronomy. Unfortunately, someone does this job already. What do I do?

Anon.
Eastbourne, Ontario

Dear Anon.,

Sorry Peter (oops, I mean, 'distressed reader'), but there are quite a few waiting in the wings, and as you well know I am already ahead of you in this queue. You'll just have to wait twenty years or so. However, I believe there are some out there trying to make this queue shorter. Only last night, I bent down to pick up a dropped eyepiece when a bullet whizzed past my ear and removed half a brick from the wall behind me. Luckily, the eyepiece wasn't broken.

Sadly, TV astronomy rarely comes high on a station's priorities. Notable trailblazers include NBC's "The Nature of Things" (1948–1954). Most TV companies, however, prefer to commission individual "specials" as the requirement arises. A continuous series on astronomy is nowadays almost unknown, although BBC's "Sky at Night" (1957–) remains a remarkable exception.

The "Sky at Night" celebrated its 50th year of transmission in 2007, having comfortably passed the milestone of its 650th episode. It also easily holds the worldwide record of a TV program retaining the same presenter.

(SOME PROBLEMS REQUIRE NO EXPLANATION.)

Hi Stevee!

I'm Suzy. I understand you're into heavenly bodies in a really big way. Use the phone number on the photo (attached) and I promise to give you a trip around the galaxy you'll never forget!

Suzy
Abilene, Texas

Suzy,

Fare is in the post. Come next week. The wife's off shopping!

Sadly, astronomy is still seen as a somewhat geeky pastime by the general population. Its "stars" do not generally inspire the fan-based wild adulation found in, say, the music business. This is ironic, bearing in mind that astronomy (the Moon and stars) is recognized as a fairly romantic arena by those within music. Brian, May the force be with you!

(KNOWLEDGE IS POWER, SO THEY SAY, AND SO FAR AS EXTRA TERRES-TRIALS GO....)

Dear Steve,

Do you think E.T. really exists? I fret that we are alone.

Carlos
Los Angeles, California

Dear Carlos,

Where have you been all your life? Of course E.T. exists. It stands for Ephemeris Time. I suggest you pay a visit to the local library and study their one book on the subject. I didn't write it.

Ephemeris time was a time scale based on the relative motions of the Earth-Moon system and the Sun, much employed by that great publication, the *Astronomical Ephemeris*. Because of ET's inherent positional instabilities, the IAU (International Astronomical Union) replaced it with TDT (Terrestrial Dynamical Time) and TDB (Barycentric Dynamical Time) in 1976.

Succeeded by the *Nautical Almanac*, the *Astronomical Ephemeris* used to be a thick annual UK government publication giving details of celestial events, including stellar and planetary positions against (ephemeris) time. Previously only available through these great weighty tomes, the same information is now available from the simplest PC software, thus most likely saving 20 miles^2 of Amazon rainforest every year. On one occasion, by coincidence the first year that this previously British publication was co-published with the US authorities, lamentable printing delays caused it to be issued very late – well into the year that it was supposed to cover. This caused considerable consternation to professional and amateur astronomers alike. When it finally became available, some quipped that it should be re-catalogued as a history book.

Fig 7.1 The sky – in a book [Image by the author]

(SOME ASTRONOMICAL AUTHORS MAKE IT LOOK DECEPTIVELY EASY TO WRITE ASTRONOMY BOOKS.)

By email
From: *Martin Puritan[mpiton@globular.com]*
To: *Dear Steve[doctorsteve@help.com]*
Subject: *Publish*

How may I become a successful author in the field of astronomy?

From: doctorsteve@help.com
To: Martin Puritan[mpiton@globular.com]
Subject: Publish

Hi Martin!

This is fairly simple. First, never write a book containing more than seven chapters. Five can be copied out of any textbook, leaving only two chapters to work on. The Introduction (a description of the material you have copied) and an epilog (a discussion on how interesting you found the material you copied) complete the work. Looking at the excellent format of your letter, this should present you no problems.

To potential authors I will only say that persistence and belief in your work pays off – in the end!

(OK. I ADMIT IT. AS A COMMUNITY, WE ASTRONOMERS HAVE MORE THAN OUR FAIR SHARE OF LOONIES...)

Dear Steve,

Science has taught us many things about the universe, much of it strange – meteor holes, black holes, wormholes, ars..whoops! (Sorry about that.) People are led to confidently comprehend time warps, infinite gravity, anti-matter, and the like. So why don't people believe me when I assert that there are fairies living in the bottom of my telescope tube? They are very sweet. And considerate. They obligingly leave the instrument just before I want to use it and then creep back into it a few minutes after I have put it back in the shed. Because they are so polite I don't ever see them. But I know they are there. They even keep the mirror clean for me (though I lightly wipe it myself occasionally to help them out). I do not understand why fellow astronomers

*think I am peculiar. Surely, you must have come across fairies that have an
interest in astronomy?*

Jonas
Bismark, North Dakota

Dear Jonas,

I certainly have. And they may be attracted to astronomers; for I have my
own, you know – though they are not always as helpful as yours seem to
be. Once, when I was doing a weekend astronomy course at a residential
college, I briefly left a full glass of whiskey perched behind me on the bar -
but when I turned to take it, the pesky little buggers had drunk it up
completely! That experience scarred me for life. I never let go of a glass
now, not for an instant.

Never let it be forgotten that in Victorian times the existence of fairies was
considered credible enough to warrant serious investigation. Sherlock Holmes
creator Sir Arthur Conan Doyle was a notable protagonist (although applying
the methods of his famous fictional sleuth should surely have put him right).
Critical examination of fairy phenomena revealed that evidence was either
mischievously conjured or lacking and the idea was dismissed. Science did its
job.

But considering fantastical ideas is the necessary meat and vegetables of
science. The mind has to be kept well and truly open. Who, not too many years
ago, would have dared to suggest that Jupiter's moon Io had volcanoes, Pluto was a
multiple body, or Triton had methane lakes! Let's just call those who push at the
boundaries of credibility "independent thinkers."

Such a thinker was born in Poland in the year 1473. By 1514, Nicolas
Copernicus was sufficiently struck by a weird idea that he intrepidly published
it in a small pamphlet (his "Commentariolus"), which he circulated to his friends
only. It was revolutionary. Literally. Instead, as everyone knew, of the Sun
circling Earth as Ptolemy had said, he suggested it was the other way around!
So radical was this strange idea that he spent a lifetime resisting the urge to
publish it properly. It was not until shortly before his death in 1543 that "*De
revolutionibus orbium coelestium*" (*On the Revolutions of the Heavenly Spheres*)
was published.

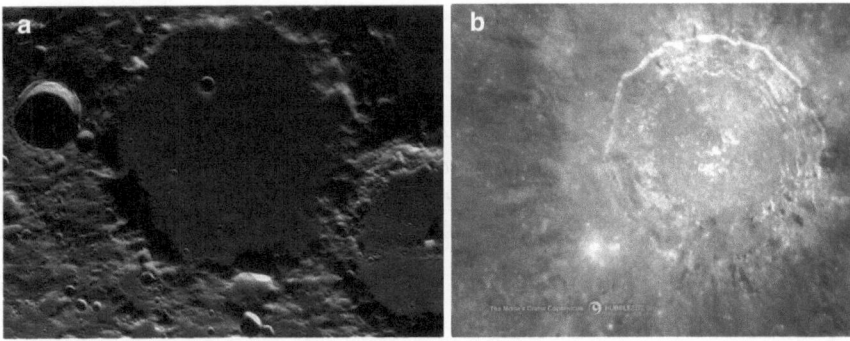

Fig 7.2 Commemorated on the Moon, a) Ptolemy's crater at 93 miles across is almost twice as big as b) Copernicus. Doesn't seem fair to the man who got it right [Images courtesy of Alan Friedman, NASA.images.org. and Hubble Space Telescope Science Institute]

It had long been suspected that Copernicus remains had been placed somewhere under the floors of Frombork Cathedral, Poland, but no one knew exactly where. Using sophisticated scanning equipment, archaeologists led by Jerzy Gąssowski found a partial skeleton in August 2005. This was comprised of some bones and a skull – although, oddly, the lower jaw was missing. Using a forensic expert to reconstruct the skull's features, the resulting face was declared to so closely resemble a self-portrait that little room was left for doubt. The estimated skull's age at death matched correctly that of a 70-year old. A scar and broken nose shown in the portrait were also matched to the skull. The clincher was a DNA match between the skeleton's bones and that of hair retrieved from a book known to have been owned by Copernicus.

(SOME ASTRONOMICAL ENDEAVORS REQUIRE EXCEPTIONAL SKILLS.)

By email
From: *Hiram Kizorsky[hysky@globular.com]*
To: *Dear Steve[doctorsteve@help.com]*
Subject: *High Noon*

Dear Steve,

For heaven's sake, you must help me, quickly. Please. Please. I am in real trouble. There was this guy I met here at the Texas Star Party. He seemed kinda normal at first, and we fell into discussing who the best astronomer in history was. I was plumping for Galileo or Newton, but this man would have none of it and insisted it was Stephen Hawking. Our discussion got heated and somehow turned into an argument. It was a bad scene, man. Suddenly he says we're going to settle this man to man and challenges me to a duel. Mano-a-mano. Tête-à-tête. I say don't be stupid, what kinda duel? He says be back here in five days at 12.00 noon, and it'll be eyepieces at fifty yards. He had a mad glint in his eye so I know he means it.

What's really spooked me is that I been asking around and found out he is a really mean hombre. Not only that, but apparently he's a dead shot with his eyepieces. Another guy I spoke to says he's seen him drop a buzzard out of the sky with a two-fingered hooked throw of a 14 mm orthoscopic. He's gonna get me right between the eyes – no kidding. Thing is, I'm not a man to back down – and the fifth day hence is the day before the star party ends. So I'm stuck here and have to go through with it. What can I do? If I go home early I'll lose face. If I stay I'm gonna get a dent in my forehead.

From: doctorsteve@help.com
To: Hiram Kizorsky[hysky@globular.com]
Subject: High Noon

I say, this does sound exciting doesn't it? Looks to me that you have no choice but to spend your remaining time at the star party practicing your eyepiece throw instead of observing. Choice of eyepiece is critical. Aerodynamics is all. A short, squat Plossl design will be better than a long Hygenian or multi-element configuration. Don't go for weight – a heavy eyepiece might do more damage, but it'll take twice the time to get to its target.

You have no time to perfect the high-velocity two-fingered pitch, but try a simple overarm, with the forefinger impelling the force via pressure against the barrel aperture. If this is a fast-draw competition, you can also try greasing the foam in your eyepiece box to make your ammunition slip out that much quicker. OK, it is cheating, but it sounds like you will need every advantage you can get. You don't mention this fella's name. Would it be Dead-Eye Dan, the fastest Plossl in the West? If so, err...good luck. When you get to hospital, do let me know how you got on. If you can.

I cannot let pass mention of Professor Stephen Hawking without describing my own singular but ignoble meeting with the great man. A few years ago, searching for presents shortly before Christmas, I was browsing the bookshelves of one of the more celebrated bookstores in Cambridge (UK). My gaze lost for a while among the spines of the weighty tomes before my eyes, I decided to move on. I turned on my heels to walk off – only to find that during my reverie a certain famous wheelchair had silently crept up behind me. Pinioned between the base of the bookshelf and the footplate of the wheelchair, my feet had nowhere to go. Unfortunately, the rest of my body was already imbued with forward momentum. Hopelessly grasping thin air to save my balance, my inevitable fall through Professor Hawking's event horizon was broken only by my clattering impact against him, the wheelchair, and his speech machine. In a clumsy heap, our eyes met briefly, just inches apart, in a mutual saucer-eyed gaze of shock and horror. Quickly righting myself, I mumbled an apology and limped quickly away from the scene in an apoplexy of embarrassment. I strained my ears for an expected staccato oath from the famous electronic voice-box, but nothing came. Edited, Perhaps, by the software the probability was that I had irrevocably damaged the speech machine and silenced the good professor forever.

If my apology at the time seemed inadequate, let me set the record straight right now and offer it to him again unambiguously (although parking right behind me in the store wasn't as entirely clever as we are led to believe, either). His work and lecturing continued afterwards – so I must assume there was a full recovery from the trauma. For me, the frozen sweaty nightmare plays out every night. But how fitting that our meeting involved a lost battle with gravity.

The postscript to this story is that a couple of weeks later, again in Cambridge, I narrowly survived when I was nearly run over by someone suspiciously resembling another well-known scientist on his equally familiar bicycle. I was left in no doubt that a revenge contract had been put out on me by the local scientific community. I trod carefully there for months afterwards.

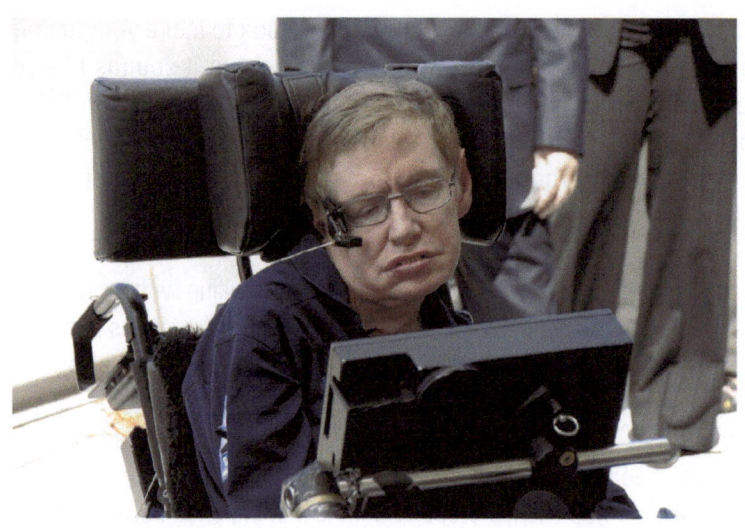

Fig 7.3 Keeping a wary eye out for me. Stephen Hawking, seen here after his NASA "vomit comet" weightless flight in 2007 [Image courtesy NASA]

(SCIENCE NEVER STANDS STILL.)

Dear Steve,

Being quite old, there was a time when, in my youth, the Solar System had eight planets. All my schoolbooks said so. I loved astronomy then. You could easily divide the planets into two equal groups; four small rocky bits near the Sun, four large gassy ones further out. Then, when I was about 14 years old, my nice neat universe fell apart. I heard that some guy called Clyde Tombaugh had discovered another one! Not only did the new small planet muck up the nice even number, it was also in the wrong place – out with the big gas giants. I was so traumatized I didn't talk for about six months, and even afterwards all I could mumble was the number eight over and over.

Psychologically I slowly recovered and got on with my life. I went to college, got a wife and kids, etc., but my reconciliation with the new addition to the Solar System was long and hard to come. However, in the last few years I had made great progress. The stammering had subsided, as had the facial ticks and twitches. Why, a few weeks ago, I was even able to pick up an astronomy book with a picture of Pluto on the front of it. My whole extended family came around to dinner to celebrate the achievement! I tell you, I was so happy that at last I had confronted Pluto and even accepted a total number of nine planets – without having a fit of the shakes. What bliss.

Then, as you must know, it happened. Following the August 2006 decision made by the General Assembly of the International Astronomical Union (IAU), there are now eight planets again. Oh, my! What a disaster! After taking 76 years to accept Pluto's discovery they've ripped it from me. My stammering has returned, and the shakes are worse than ever. I mean, it is not as if Pluto floated off somewhere of its own accord. Someone simply demoted it!

I feel I don't belong anymore. Am I in a Solar System of eight or nine planets? I no longer know where I am. How can I readjust my adjustment after all this time? My psychiatrist warns me that this time I may never recover. The pills don't work anymore. Can you help me?

Matthew
Durham, North Carolina

Dear Matthew,

Listen. Get a grip. You've had it easy. My great-great-grandfather started with seven planets and had a nervous breakdown every time they discovered a new one in the early 1800s. Don't forget, at one stage the Solar System's planet population got well into double figures. His sanity was saved only when they redesignated these new bodies as minor planets, when the total dropped back to seven again. (Unfortunately the discovery of Neptune in 1846 did him in completely, but that is a matter our family doesn't talk about!)

Also, about the speech problems and trembling. Are you sure about the reason? How can I put this? Do you think this might be a teensy-weensy bit to do with the fact you are 90!

On August 25, 2006, inhabitants of the world woke up to find themselves in an entirely different Solar System. This was no "Star Trek" transporter malfunction, but the effect of a little known IAU working group rather pompously entitled the "Planet

Definition Committee". Its mission had been initiated by the embarrassing surplus of Pluto-like bodies being discovered at the edge of the Solar System. Were they planets, on equal footing with Pluto? Or were they and Pluto bloated members of the Kuiper Belt, hitherto merely considered a distant cousin of the Asteroid Belt? Was Pluto then just a large asteroid and not a planet? These questions were giving astronomers sleepless nights. Apparently.

Working closely (and, as it proved, a little too quietly) with a small band of consultants, the Planet Definition Committee brought their solution to the 2006 IAU meeting for rubber-stamp approval. Yet unexpectedly, and despite the committee's 3 years of diligent analysis that preceded it, consensus among the assembled astronomers of what constituted a planet remained elusive. Chairs were not thrown, but only barely. Nevertheless, late in the day after some "horse-trading," certain dynamical parameters were agreed and passed. During the afternoon of August 24, 2006, Pluto found itself demoted to a new subclass of object – a dwarf planet. Nine became eight.

The storm broke the next day. The general public were aghast; the media were frenzied. Everyone knew that Pluto was a planet. It had been for 76 years, so how could it now not be? Scientists, too, were raging. The hurriedly corrected redefinitions agreed at the assembly, they said, were inadequate or downright wrong. Dynamical astronomers and planetary geologists, seeing the problem from different

Fig. 7.4 Pluto and its "moons" [Image courtesy of A. Stern (SwRI) and Z. Levay (STScI) NASA.images.org./ESA]

perspectives, began a blistering, undignified, and very public argument about the definitions. Harsh words were said. And most seriously, why hadn't the whole international scientific community been consulted? After all, the change had been defined and "authorized" only by a handful of those astronomers at the assembly (and then only after many members had left to return home). Compared to the number of professional astronomers worldwide, this radical motion had been ratified by a very small representation indeed. This was its greatest weakness. Justly or not, the IAU found itself in hot water and the unexpected center of international criticism.

Unhappily, in retrospect, it was not handled well. The result is a continuing and occasionally intemperate debate within and outside the astronomical community. New, increasingly desperate proposals for planetary criteria are hatched almost daily. Sadly, the unedifying furor diverts attention away from the real drama, that of discovering fascinating new worlds at the cold distant edge of our Solar System.

(SOME PROBLEMS ARE TOO CLOSE TO HOME BY FAR.)

Dear Steve,

I am not going to give you my name, for obvious reasons. But I need help about how to deal with my employer. I am a sort of assistant to someone who is in 'the astronomy business.' My problem is that he is a really mean bastard who doesn't pay me properly for the work I do. In fact, most of the credit he gets in sorting out various astronomical problems is due to me for my own hard unacknowledged background work. I have asked him for an increase in salary for three years running now, but he just gives me the brush off. He says he has a large number of 'hidden' costs; but so far as I can see, these are like dark matter – can't be seen and can't be specified. How do I deal with this stingy egotistical tight-fisted toad?

Anon.
Boston, Massachusetts

Hi, Mr. Anonymous,

I know what's up! If you think I should give you extra money just because you have to open an increasing number of envelopes and e-mails each week you must be joking. I can't stop you reading them, of course. And maybe I do occasionally translate your pithy and frequently relevant sage-like comments into my replies. Okay, perhaps even 'occasionally' is a little ungenerous. Will 'frequently' do? But even if I use your input most of the time, is it MY fault that you are willing to then type these out for the Dear Steve column under my name? Yes, I know that you would not be doing all the work if it wasn't for those special pictures we often talk about. But then, you maybe shouldn't have gotten drunk that night when I took them. I mean, I was hardly going to resist recording the multiple exposures I made when you were hanging by your fingernails from the top end of the Yerkes 40 inch refractor. Don't blame me. The coffee evening after the Astronomy Symposium that year at Chicago University was supposed to be alcohol free – it wasn't ME that secreted a quart of whiskey in your back pocket. How you drove us to their observatory at Wisconsin Bay I'll never know. Of course, the forced entry was down to you, too – but we won't mention that either, huh?

I've got all the photos in a safe place. Remember that! Until I have decided you are no more use to me than a drunken horse, you will do all my bidding, all my work, and get a big fat zero in your paycheck and effort recognition t'boot. And, of course, it's entirely down to you to ensure this letter never makes it into the printed problems. Right?

Sadly, although there are a few high flyers who command big bucks, the astronomical community at large is paid rather poorly. Indeed, science generally is seen by private and public institutions as a vocational endeavor, within which the reward of scientific progress is viewed as a significant but unspecified portion of salaried remuneration. This trend changes, of course, depending on the culture within which our fated (perhaps even feted) astronomers find themselves. By their arcane knowledge, astronomers have been in turn both servants and masters of their communities – with salaries concordant with that status. In Victorian Britain, the rising popularity of astronomical interest meant that each great estate with its country house had an observatory built and installed with its own "pet" astronomer. The modus operandi was for the ragged astronomer to spend sleepless nights making observations, while the gentrified employer took all the credit for their published results.

At other times, in different places, those with a knowledge of the sky and its seasons commanded great influence by their needful guidance of agricultural civilizations. Indeed, the need for such skill would have been coincident with the first early human settlements. In the fertile region between the Euphrates and the Tigris (the ancient region called Mesopotamia), the Sumerians of 5,000 years ago were already developing an astronomical awareness. It was they who initiated the sexagesimal system (base 60) that was passed to our intellectual forebears, the Greeks, through the Babylonian civilization – thence becoming the foundation for our modern measurement of the sky and time itself.

(THE PROGRESS OF SCIENCE IS NOT SMOOTH. IT IS LUMPY.)

Dear Steve,

We need you to prevent a murder, because unless you sort this argument out one of us is going to kill the other!

I'm Mark – this is my bit.

It's this business about the Oort Cloud! As you obviously know, this is a proposed exterior region of the solar system containing the debris left over after its formation – supposedly the source of mountainous dirty ice balls sometimes perturbed towards the Sun and the inner planets as comets. I say 'supposedly' because, as I say to my stupid colleague, Alan (another member of our astronomical society), this cloud has never been observed and is only inferred by the orbital parameters of long-period comets. It is a total fabrication by theoretical astronomers and hence a load of rubbish! Not one single teensy-weensy bit of comet has been detected in this region. In fact, what IS being discovered 'out there' are fairly large asteroid-like bodies resembling the dwarf planets Ceres and Vesta!

My gullible colleague is adamant that the existence of the Oort Cloud is not disputed and its direct observational discovery is only a matter of time. This is plainly wrong, but Alan won't accept it. We used to be friends, but his stubborn intransigence on this issue has revealed him to be a willing ignoramus of the first water, and our continued arguments on this are now disrupting our astro group meetings. Last week we were both told to leave the room during a general discussion about astronomical artwork, as we still managed to guide the subject on to the Oort Cloud. (I think I said something about it being the one thing in the Solar System you couldn't draw, even imaginatively.) That said, we were off! Blows were exchanged.

This is Alan's contribution:

Hi, Steve. Mark has, unfortunately, become rather fired up about this issue. It must be said that he has not been doing astronomy for very long and so remains pretty thick and stupid on many issues – including this one. He is unaware of the vast amount of literature lending weight to the Oort Cloud's existence. Indeed, I don't believe he does much reading at all; watching TV documentaries is about his deepest level of study, sadly. He is a walking testament to my view that prospective members of our group should pass an entry exam. We really don't need self-opinionated fools like this around. Our chairman, in desperation, has suggested we write to you in order to settle the issue once and for all, with the loser undertaking to abide by your decision and leave our fold forever. So far as I am concerned, the sooner Mark leaves the better, especially as my girlfriend has already said that he is better looking than I am!

Mark/Alan
Ogden, Utah

Dear Mark and Alan,

What a sad pair you are. You really Oort not to be arguing like this! (Tee hee, my little joke! Sorry.) You are actually both correct. It is true that conjectures on the origin of cometary material lean very strongly towards a distant donut-shaped expanse of such distributed material at the limits of the Solar System. Orbital characteristics indicate this. However, our current technology does not permit direct detection of it – a problem you may appreciate when you consider that these perhaps 10-mile-wide objects may be more than 50,000 times further from the Sun than we are!

Curiously, my postulation that you are both correct perversely means that you are also both wrong. This being so and in consideration of your assertion that the loser of the argument must quit membership of your society, I regrettably must decide that BOTH of you must leave. This is my final and absolute judgment. Of course, in the future, should this matter be unambiguously settled, one of you may return. But it may be some time.

It is perhaps a general perception that scientific disputes are settled in a gentlemanly way, with polite discourse taking place in a genial deferential arena. This is *of course* largely true, although those of us who have been "in the business" for long enough have certainly witnessed some tumbles from this noble ideal. Such failures of good humor normally occur behind closed doors, between groups with vested interests. Occasionally, though, scientific antagonisms do expand into the public.

One of the most famous incidents must be the verbal exchange, during the British Association's Oxford meeting of June 30, 1860, between "Darwin's Bulldog" Thomas Huxley and Bishop Samuel Wilberforce. The latter had sneeringly asked Huxley whether it was through his grandfather or his grandmother that he claimed descent from a monkey. Huxley retorted that although he would not be ashamed to have a monkey for his ancestor, he would of a man [like Wilberforce?] who used great gifts to obscure the truth. Such was the understated Victorian vigor of this riposte that the attending Lady Jane Brewster (second wife of scientist Sir David) gained eternal scientific fame by fainting from shock. One can just imagine the consequent scene, the hall suddenly reverberating to astonished murmurs at the swift intemperate tone while a clutch of concerned nearby audience members rub Lady Brewster's hands to restore her prone body to consciousness. I would really dearly love to have been there! (Such is my decrepitude that there are some who contend that I was!)

Thankfully, astronomy has its own grand historical disputes. One of these occurred in April 1920 (by pre-arrangement!) at a meeting of the National Academy of Sciences in Washington. By the early twentieth century, astronomy was wrestling with the vexed question concerning the nature of the so-called spiral nebulae, so famously observed in the mid-1850s by the third Earl of Rosse through his 72" reflector. There remained two schools of thought on what they were. Either our "universe" consisted of our galaxy of stars and the spiral nebulae were smaller objects of unknown nature embedded within it, or our galaxy was an island within a grander largely empty universe, and the spiral nebulae were other galactic "islands" at great distances. Since in the early 1900s no stars had been observed in these nebulae and their distance had yet to be gauged, this was still a mystery.

While the prosaic preparations of the April meeting were being made, it was felt that a "debate" should be arranged to elicit some interest – dare we say, even excitement. Discussion on which subject should be chosen for this debate eventually narrowed down to either relativity (a relatively [!] new subject), or on the nature of the enigmatic island universes (spiral nebulae), upon which no less than the scale of the universe was hinged. The latter subject won the day, as it was deemed sensible that "more than half a dozen members" should understand what was being discussed!

Eventually, it was arranged that Heber Curtis of Lick Observatory would speak on behalf of spiral nebulae being independent bodies exterior to our own galaxy while Harlow Shapley would supply the counter argument of gaseous masses within our own all-encompassing galactic universe. Apparently, Curtis became keener for the debate as its date approached, having the advantage of being an accomplished public speaker. Shapley, on the other hand, was at the time trying to avoid making waves (for pivotal career considerations) and came to the table much more reluctantly. Looking back on this encounter, it is hard not to get the distinct impression that both protagonists were pawns being maneuvered into a pre-arranged scrap. Certainly, others involved seem to have been rubbing their hands with glee at the prospect of an encounter to remember. No boxing promoter could have done a better job at engineering such a confrontation!

It is hard now for us to understand that both men were in weak positions; distance measurement at the time did not permit accurate assessment of stellar scale – that of our galactic neighborhood or the ambiguous spiral nebulae. Each astronomer was to be given 40 min to state his case, after which an open "conversazione" would take place. Thus, at 8:15 on the evening of April 26, 1920, the battle was joined. Shapley opened "the discussion" with his paper on spiral nebulae being members of our own galaxy of stars. Curtis followed with his superior delivery, espousing the distant, independent option. As Shapley ruefully recalled later, it was the smoother performance of Curtis that won the day. However, each followed up the meeting with published papers in the following May that show both opinions were equally plausible and well argued. Both corners came out of the fight with honor, since the exchange of views galvanized the topic and initiated a fresh impetus for its solution.

Although our modern view of the "Great Debate" is often collapsed into the exchange at this meeting, it more properly embraces the papers subsequently written by the protagonists and the wider scientific scrutiny that followed. I get the impression that although much was at stake at this meeting, both scientifically and personally, it was nevertheless a dignified, honorable, and sedate discussion. Maybe it should be called instead "The Great Sedate Debate"?

The question remained fully open until Edwin Hubble burst on the scene, in almost theatrical fashion from stage left, clutching his discovery of Cepheid variables in the Andromeda "Nebula" (M31). It was these celestial standard candles that allowed a distance measurement revealing that the Andromeda Nebula was in fact not a denizen of our extended star system but an independent island universe outside our own – a galaxy in its own right. Nevertheless, despite the passage of 80 years since, one can still find contemporary inappropriate references to the Andromeda Galaxy as the Andromeda Nebula!

As an interesting postscript to this story, it is ironic to discover that the galaxy that first established its great distance from us is actually on a collision course with our own, closing in on the Milky Way at an astonishing 120 km/s. The Andromeda Galaxy arrives at our region of space in 2 billion years time, with subsequent gravitational entanglement that will result in a shotgun marriage 3 billion years after that. You have time to pack. [Debate info. from The 'Great Debate': What Really Happened. Michael Hoskin, J. Hist. Astron. 7, 169-182.]

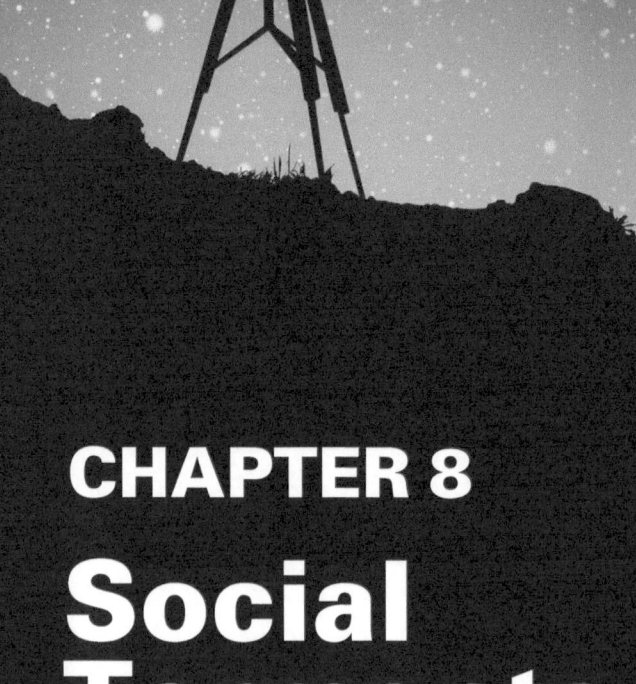

CHAPTER 8
Social
Torments

" Aaah ! He's an astronomer ! "

By its nature, observation of the sky is a singularly personal experience. One eye is pressed up to the eyepiece of a telescope, yielding a unique view on the universe. Indeed, some say that it can never be known for sure that what one observer sees is exactly the same as that seen by another. But there is no room for that sort of philosophizing twaddle here! Astronomical observation needs no crowd of adoring onlookers, no teammates to pass the ball, no steadying hand of a wise sage, no dancing line of rah-rah girls, or group of supporting technicians. Communion with the cosmos requires but a lone observer. And this is how most astronomers start out — on their own.

As a consequence, it might be suggested by perceptive people that astronomy is an attractive pastime for those who like this state of being and prefer their own company, spurning the cluttered complications of social interaction. But in the end the desire to share the pleasures of celestial ecstasy becomes too great, and occasionally astronomers like to come together. One is reminded of other species, such as bears, salmon, and turtles, living solitary lives except for occasional conglomerations to breed. Yes, astronomers are like this, too. Yet the consequential discordant clash of simultaneously endeavoring to be part of a group and steadfastly maintaining an

S. Ringwood, *Astronomers Anonymous*, DOI 10.1007/978-1-4419-5817-4_8,
© Springer Science+Business Media, LLC 2010

observational independence gives rise to unhappy stresses. This can be a particularly tricky problem if a committed heavenly passion is to be inflicted on a lifetime partner who does not share it.

Alas, the adoration of celestial pinpricks in the middle of the night gives the eager observer a reputation for eccentricity. Some simply endure it, while others positively cultivate it. Certain individuals who shall remain nameless come to mind. In all events it does imbue a sense of otherness.

Come then, with me, to share our pain.

(NOT EVERYONE IS GUARANTEED TO BE SUPPORTIVE OF AN ASTRONOMER'S NOCTURNAL ACTIVITIES.)

Dear Steve,

I am not an astronomer. But I have the misfortune to live next door to one who is always up and about in his backyard at all hours of the night crashing and grunting about in the undergrowth with a long tube and a small red torch generally making awful and unwelcome "Aaah!" noises that keep myself and my wife awake at night. Can you suggest a way of asking him to desist?

Anon.
Boston, Massachusetts

Dear Harold,

If astronomers want to use their telescopes late at night it's your very bad luck! You shouldn't have ears like radio dishes! And *please* stop writing to me at this address. Why waste stamps when you only live next door.

I am afraid that as an amateur astronomer you cannot automatically expect those around you to empathize with or indulge in your nocturnal activities. Hard as it might seem to imagine, there are many who are simply not interested in astronomy and will take a dim view of your nighttime goings on. Consideration cuts both ways, so if you need a neighbor to trim a tree or curtail the use of a security light, repay that favor by keeping the noise down at 2:00 in the morning.

Fig **8.1** Radio telescope at Cambridge, UK [Image by the author]

Astronomy has stranger enemies than a disinterested, nay antagonistic, neighbor. Drivers traveling along the M11 highway near Cambridge, England, may wonder why a solid metallic barrier adorns one side of it. This is to prevent the sensitive detectors of the nearby world-class radio telescopes from being deafened by the radio-noisy electrical systems of the cars speeding by. However, a confusion of sources between a distant exploding galaxy and the radiative rattle of an old car does seem unlikely.

(CONFLICTS OF INTEREST ARE A CONSTANT SOURCE OF TORMENT.)

Dear Steve,

My wife seems very jealous of the time I spend in the backyard observing with my telescope. Apparently, I pay more attention to my telescope when there is a clear sky than to her winsome temptations in skimpy lingerie. This is in spite of the hastily scribbled messages of affection that I leave her each time I begin my long observing sessions. I do not see that she has cause for complaint. Please advise – my marriage depends on it!

Daniel
Fresno, California

Dear Daniel,

Your loved one *must* come first. I fear that divorce is the only answer.

As the noted UK Oxford University luminary Dr. Allan Chapman puts it, "Amateur astronomer" describes someone who does it simply for the sake of love (from the Latin *amata*). Ironic, really, since it frequently gets in the way of it.

Some amateur astronomers (such as me) are lucky enough to have a partner who shares their interest. For others, a spouse's gritted indulgence is required. To those latter unfortunates let me recommend that a little consideration goes a long way; like warming the feet up *before* returning to bed after a winter night's observation.

Some wives are pure inspiration; take the story of Asaph Hall. In 1877, using the US Naval Observatory's 26″ refractor (then the largest refractor in the world), Hall was trying to find out if Mars had any satellites. After a long and fruitless search he had all but given up, concluding that Mars moved alone. His wife of 21 years, Chloe Angeline Stickney, pressed him to continue for one final effort. (It could have been that after some blissful solitude, she did not want him under her feet.) It was during this last throw of the dice that he discovered Deimos on August 12, then Phobos 6 days later. Chloe Stickney's contribution is commemorated on Phobos by a crater of that name.

I should add hurriedly, lest the impression is given that Chloe was simply offering tea and encouragement, that prior to their marriage she was *his* professor of geometry at McGrawville Central College. She helped him in his work, but did less of it when he refused to pay her a proper wage for the job!

Fig 8.2 Scope or the wife. . . you decide [Image by the author]

(BODY ODOR [B.O.] WAS A TERM INTRODUCED BY INVENTIVE DEODORANT ADVERTISERS. BUT ASTRONOMERS CAN SUFFER FROM MORE THAN A BAD SMELL!)

Dear Steve,

I am so embarrassed. Last week during a meeting of my local astro society a friend leaned towards me and whispered that my telescope had B.O. I did not know where to look! I had not even realized. I was advised that it needed pollinating or

*something. I caught some bees in the backyard and stuffed the telescope with them
for three days. Not sure if there is any difference. What should I do now?*

Adam
Memphis, Tennessee

Dear Adam,

You moron! They were telling you that your telescope has <u>B</u>ent <u>O</u>ptics
and needs *COLLIMATING.* (This does *not* mean a sexual liaison with a
cauliflower!) The optics need to be realigned by an experienced person
who knows how to use the sophisticated tools involved. It is a delicate
operation requiring the deft use of a chisel, hammer, crowbar, and bolt-
splitter. Be on hand so that you can learn to do it yourself next time it is
required.

For telescopes to be at their best, the optics need to be aligned with each other
perfectly. Should the optical axis of even a single component be so much as a trifle
out of kilter, the performance of the instrument will be degraded significantly. The
effect of misalignment becomes worse the lower focal ratio you go. Fortunately,
there are devices (usually inserted at the eyepiece port) that make this task (called
collimation) quite straightforward.

The best known of these is called the Cheshire eyepiece. These are freely
available commercially. Current technology now features the use of lasers in this
task, too. All telescopes should be checked frequently to ensure optimum performance. With my tongue firmly in my cheek, I tell you that it can be an activity that
takes up more time than actual observations made after its completion.

Collimation of a telescope is akin to the tuning of a piano. A top-quality performance
in both cases can only be ensured by a very fine and accurate adjustment. How delicate
and refined these adjustments need to be can be underlined by reminding ourselves that
we are dealing here with wavelengths of light. Even a slightly misaligned telescope will
induce effects that are comparable to the defocusing of an image.

All moderate to large telescopes should incorporate a feature that allows the
tilting of optical components. Claims from any manufacturer or retailer that this is not
necessary should be viewed with suspicion. Simply moving a telescope around can
impart motions sufficient to move the optical axis of a component by a detrimental
degree.

It is most common to see such adjustments supplied by push-pull screws on three
axes, both for mirror and refractor lens cells. With consequent adjustments available in

three axes, plus the well-noted complaint that the direction of these adjustments often seem counterintuitive, the exercise of collimation can be, shall we say, stressful.

(IT'S NEVER AGAIN AS GOOD AS THE VERY FIRST TIME.)

Dear Steve,

Yes! Yes! Yes! I finally did it, last night. As an 'innocently delicate' properly brought up young lady, I know that I am supposed to wait until I'm old enough to know what I am doing; though I'm not that young. It's not as if it's terribly wrong is it? After all, if the will is there, shouldn't we follow our natural urges and instincts? I'm not ashamed of admitting it anyhow. I've just done it for the very first time. I'm not ashamed to admit that it was wonderful. It made me warm deep down inside, especially as my boyfriend helped me. What the hell, you have to take your very first look through a telescope some time! The only thing is; how do I tell my mom? She's very middle class, knows nothing about astronomy, and would be shocked to discover that I had now at last looked properly up at the sky.

Tiffany
Muncie, Indiana

Dear Tiffany,

I understand your concern. I waited ten years before I admitted as much to my own mother – and it was she who bought me the telescope in the first place! Take your time and do it gently. You have to understand that your mom sees the physical act of your observation as indicative of your growing maturity. This symptom of your aging serves only to remind her of her own. Explain that the temptation was too much for an intelligent soul like you. Innocence may have gone, never to be regained, but the expanded universe that now lies before you is worth the sacrifice. It helps if you ask her along to be there next time you do it. But why watch - she may get the chance to enjoy it too! If you think it would help I would certainly like to offer my services in furthering your nocturnal development. . ... and that of your mother.

No-one forgets their first look through a telescope. My "first light" took place during a family holiday in Jersey (a British island off the French coast) when I was 15. Having saved up a (then!) princely sum of over £7, I was quite unexpectedly able to buy a small 40 mm refractor – despite this purchase exhausting my spending power for the remainder of the holiday. Setting up that night on the terrace of the hotel bar, the thrill of those tabletop observations of the first-quarter Moon stays with me still.

Fig 8.3 Saturn. No one forgets their first time [Image courtesy of Jeff Barton and Josh Walawender, Saturn Observation Campaign/NASA.images.org.]

(ASTRONOMICAL ACTIVITIES CAN APPEAR ALARMING TO THOSE OUTSIDE ITS CLOSETED ENVIRONMENT.)

Dear Steve,

I know that a mother always worries about her children, but I desperately need your advice. My lovely, charming, innocent daughter has recently taken up with a bunch of oddballs who claim to be amateur astronomers. She goes out in the evening saying she's going to look at the planets and returns at all hours of the night praising someone's 10" instrument. When I demand to know what she has been doing I am horrified. Apparently she has been observing loony occultations – which sounds absolutely disgusting to me – and dusky umbrellas. Who are these people she has gotten involved with? I am so shocked; she used to be such a normal decent girl. It's gotten worse lately. She told me one of the guys let her use a retractor attached to his Dobbo! What am I to do?

Anne
Muncie, Indiana

Dear Anne,

Relax! I assure you that astronomical societies have no more deviants, murderers, religious fanatics, alcoholics, and criminals than are represented in the average population. I hope you find this reassuring.

If there's one thing that will get me ranting from the lectern it is the oft expressed view of amateur astronomers being oddballs (this book excepted!). I mean, where else will you find a caring bunch of intelligent people who actively promote science for no financial gain, use their spare time to admire the universe's handiwork, carry out fundamental scientific studies of benefit to worldwide knowledge, actively encourage education without barriers of any kind, and generally dwell on the human race's place in the universe?

Er...well, we do have a few weirdos. But we look after them. Who else would?

(IT IS A TRUISM THAT ANY GROUP OF ASTRONOMERS UNDER A CLOUDY SKY [OR EVEN A CLEAR ONE] WILL BE FOUND IN THE NEAREST BAR.)

Drear Stweve,

S'funny thwing. Is. I lye astronomonomy. Lot. Bud everydime I loof frew a williscope I canno see buggerall. Why? i goggle a' the easypiece an' screw careflee, but nufin. . . . . .all. The sky moos an' it tracks, but no speckle shows, e'en a littlebitty. Bit. (Weee, i's spinning; suffin' from preception of the proles.) Shcope's lined an' poinin', but no sparkues. Wobblies get it anit looks at leaves. An' doan look me! Should I shell it? Bruddy fing's gota hole in . . .end anyway. Stuff it. I see fine Moon my self- don' need stilly shoob. Wanna by?

Caan remmbrnum, hooooooo cares fum.

You need AA. (Astronomers Anonymous.). In the meantime, wee dram is in the mail!

I have often considered that such astronomically biased self-help organizations would be quite useful. Each time I do so, I realize that they already exist. There are branches everywhere. They are called astronomical societies!

There are approximately 300 astronomical groups in the United States, with 160 in the United Kingdom. All are direct descendants of the plethora of special interest leisure groups that were born in the Victorian era. At that time, the Industrial Revolution created for the working classes something that had hardly existed for them before. Spare time. This allowed far more people to follow individual interests – including astronomy. These individuals soon came together to form learned societies of literature, craft, and the sciences.

Belonging to a society almost became a national pastime. By the late nineteenth century, there were sufficient numbers of amateur astronomers to form, in the autumn of 1890, the UK's British Astronomical Association. On the other side of the Atlantic, less than a decade later, the American Astronomical Society was born.

(SOME, ALTHOUGH I CAN'T UNDERSTAND WHY, ARE NOT INTERESTED IN ASTRONOMY AT ALL. THIS CAN THWART ROMANCE.)

Dear Steve,

I have been going out with a girl for two years now and conversation is beginning to turn to marriage. Quite frequently. I think it's a great idea, but the trouble is, I have not yet told her about my astronomical interests, or the 200 mm reflector hidden under a tarpaulin in my backyard. I fear that once she discovers my dark secret this may put her off me. Do I tell her now, or delay it until after the marriage? She's a redhead.

Dennis
York, United Kingdom.

Fig **8.4** "Do I now tell her about my real passion?" [The author and his wife Gillian upon their marriage. Image by the author]

Dear Dennis,

This is tricky. Some girls will run a mile, whereas their partner's nocturnal activities may turn others on. You give me no clues about the girl herself, which makes it difficult for me to judge which category she would fall into. The red hair, of course, indicates caution. In the long run, it's probably better if you tell her before the wedding. She will find out soon enough anyway (although she must be pretty thick if she's missed your enthusiasm for astronomy up to now!). Be gentle. Introduce the subject by stages. I always find that a good line to start with is ". . .you are as pretty as a star. And by the way did you know that the star there above the power station is. . ." etc.

Don't, for heaven's sake, tell her straight off you are an amateur astronomer. Allow her own interest to gather some momentum first. Pretend you are an interested beginner, too. As the subject then becomes more commonplace in your conversations you can impress her with your 'newly' acquired knowledge. This can then lead to a dramatic climax, when you proudly reveal you have an 8" instrument! I guarantee this is bound to leave her gasping and panting for more. (Note: I do *so* like wedding cake!)

Lots of girls ARE interested in astronomy. I have to say that within my immediate sphere of knowledge, I know of at least six married couples who met at their local astronomical society – including my wife Gillian and I. So girls, if you are looking for a worldly, intelligent partner who has the looks of an impeccably handsome man (at least, in the dark!), bear in mind that your gender will be outnumbered at least six to one in your local group.

This is not to say that women are not cerebrally represented in astronomy. Elizabeth Hevelius made observations as a partner to her illustrious husband; Caroline Herschel (b. 1750) assisted her planet-discovering brother and discovered eight comets herself; Henrietta Leavitt (b. 1868) discovered those standard candles of cosmic distance – Cepheid variables; Jocelyn Bell Burnell (b. 1943) discovered pulsars; Carolyn Shoemaker (b. 1929) has 27 comets to her name; even Galileo's eldest daughter, Suor Marie Celeste (b. 1600), was a continuing source of encouragement and strength during her father's "difficult" times with the church.

(YOU FIND PRIMA DONNAS IN EVERY WALK OF LIFE, AND BELIEVE ME, THEY'RE OUT THERE IN ASTRONOMY TOO! WHAT FOLLOWS IS LOOSELY, BUT NOT SOOO LOOSELY, BASED ON AN ACTUAL LECTURER AND HIS

PRE-ARRIVAL DEMANDS. I WILL NOT SWELL HIS ALREADY INFLATED EGO BY IDENTIFICATION. BUT IF YOU'RE WATCHING, YES, I MEAN YOU!)

Dear Steve,

I am an amateur astronomer who lectures to local astronomy groups. At first, I seemed to be quite successful. Recently, however, I have received no invitations from anyone. I cannot fathom this out since I'm a really great lecturer. I'm so good in fact that I marvel that groups are not falling over themselves to get so much as a phone call from me (in fact, I don't charge too much for those calls). Why is it that these pathetic groups of wretches, whose intellects reach no higher than the back of my knees, do not crave the incredible honor of a visit from me? It's not as if I charge extra for my cigarette health warnings (i.e., "If someone smokes I'll stub the damn thing out on their forehead").

I also excel at improving the punctuality of astro meetings by balling out anyone arriving after I've started. If I'm fool enough to let them off being late, for one of MY lectures, they are barred from making even the teeniest noise – even to ask questions afterwards. I improve the quality of attention, as I make it clear that under no circumstances will I tolerate even the sigh of breathing above a certain level. And if, at my prior request, they haven't had time to repaint their meeting room to my stipulated color it's hardly my fault is it? I don't think it is unreason- able, either, to demand they drop to one knee when I am introduced to them. And can it be so much trouble to be granted honorary chairmanship of their pathetic group for two years afterwards? What I really can't stand is their attitude!

Frank
Norwich, United Kingdom

Dear Frank,

I feel that your obvious talents are aimed at the wrong market. I'm sending you by mail an application form for a position giving a short astronomy course at the Sans Tete Institutios del Astronomios dos Papua, New Guinea. The local tribesmen there are really keen to get ahead. Yours should do.

When lecturing to local astronomical societies, it has been my unfailing experi- ence to receive a courteous and friendly reception. This makes a return to that group more likely (at least on my part!). Astro societies should always endeavor to treat their visitors kindly if a revisit is desired. For those who don't, word gets around quickly.

By the same token, speakers should always be sympathetic to the sometimes adverse facilities and conditions in which some less than fortunate societies have to meet. Alas, I know of some contrary prima donnas possessing no such consideration whose names will not sully these pages.

(ASTRONOMERS OFTEN INDULGE THE PUBLIC'S LOVE OF ECCENTRICS.)

By email
From: Emily Danton[danem@globular.com]
To: Dear Steve[doctorsteve@help.com]
Subject: Alone

Do you have relatives who are interested in astronomy, too? I get so depressed by my family's dismissal of my hobby.

From: doctorsteve@help.com
To: Emily Danton[danem@globular.com]
Subject: Alone

Hi Emily

Alas, I am an only child. I'm afraid, too, that I have precious few relatives who even look up from their TV, let alone at the sky! One or two of them used to be interested in astronomy, but when I began to be interested they gave it up abruptly. Can't think why!

However, all is not lost. I can claim with some justification that those I have enthused to take up the interest are related to me as kindred spirits, spawn (as it were) of my astronomical seed cast to the four corners of the universe. Quite a few of them make a good living. We are taking over the world, you know. . .

When introduced to new people with the tag of their interest, astronomers everywhere are used to the same response. The exclamation "Really! How interesting!" is swiftly followed by a step backwards and a sudden desire to find the bathroom.

I am now quite used to being regarded as a little bit strange, simply because I like to stay up all night looking at distant specks of light. I am experienced now. I wait until someone is boxed into a tight corner before I tell them I am an astronomer.

But even the strangest of current day astronomical eccentrics have a hard act to follow in Tycho Brahe, the seventeenth-century Danish nobleman (already mentioned elsewhere for his observational acumen) who became the court astrologer to King Frederick II of Denmark.

Fig 8.5 The noble Tycho [Image from the *Astronomiae Instauratae Mechanica*, 1598]

His first claim to fame was getting his nose cut off in a duel with a fellow student over a point of mathematics – although they remained friends. (I'm tempted to observe that his opponent "won by a nose," but for the sake of my readers' sanity I won't.) Once granted his little observational fiefdom of the island of Hven, he ensured that his new "palace of the heavens" incorporated a jail so that he could properly punish wayward serfs within his domain. To keep him company during his astronomical endeavors, he had two pets. One of them, an elk, lived with him in his house; this companion unfortunately died through falling down a flight of steps because it was drunk, having tanked up from a barrel of beer on one of the upper floors. The other pet, if he could be called such, was Jepp the psychic dwarf.

Despite these lordly insanities he did sterling work with his sighting instruments, his observations eventually being used by Johannes Kepler for his historic discoveries on planetary motion.

Tycho even managed to die in style. A heroic eater and drinker (gastronomer as well astronomer!) he was dining one evening in October of 1601 at a state banquet given in Prague by Baron von Rosenburg. (Tycho had only 2 years previously moved to the welcoming patronage of the Hapsburg Emperor Rudolf II, fleeing the antagonism of Denmark's new King Christian IV.) Towards the end of the meal, despite being desperate to go to the toilet, he felt by etiquette he could not do so before the attending nobles had left the table. He apparently held nature at bay until his return home that evening, yet then with horror found himself unable to relieve his bladder. Following a subsequent rupture of his digestive system, he died of urinary infection 11 days later.

At least, that is how this story has come down to us through the centuries. But possibly, history is about to be changed. The background to this new development is that strands of Tycho's beard were analyzed in 1991 and 1996, revealing a large ingestion of mercury only hours before he died.

The symptoms of his death, as related by contemporaries, do match those of poisoning by mercury. Until now this discovery had left two possibilities open. A likely though tragic explanation is that Tycho mistakenly killed himself with a then conventional but inappropriate dosage of mercury to treat his uremia (from which ironically he may have been recovering). A newer, darker theory has suggested that the administration of a *second* dose of mercury was a murderous coup de grace, a lethal follow-up to an initial dose delivered secretly at the banquet.

Two suspects who had motive and opportunity are in the frame. The first is Johannes Kepler – an ambitious colleague frustrated by Tycho's reluctance to release to him desperately needed observations. The second is Count Erik Brahe, a cousin of Tycho's. Recent research alleges that this distant relative carried out the deed under the instructions of Tycho's previous employer, the vengeful Danish ruler King Christian IV (although whether for personal or political reasons is unclear). A murder of Tycho would certainly be in keeping with his colorful lifestyle!

So, was his death benign or malicious? The truth may at last be revealed. In an exciting proposal led by Danish archaeologist Jens Vellev, a request has recently been submitted to exhume Tycho for further tissue analysis – although at the time of writing no date for this has been set. Four hundred years after his death, Tycho still courts controversy and intrigue.

(WE ASTRONOMERS ARE A PERSECUTED LOT!)

Dear Steve,

I have neighbors who think I am strange. This is simply because I repeatedly go out into the cold night air to observe, through a tube, the indistinct blobs of light that hardly change night after night. Should I have the misfortune to be spotted in the backyard, I have to endure a rising cacophony of catcalls, derision, and raspberry blowing as neighbors call upon each other to join in. It doesn't matter how discreet and quiet I am, there's always one of them on watch to spot my nocturnal activity. What can I do, apart from move?

Troy
Alice Springs, Northern Territory

Dear Troy,

I fear that, generally, people will always try to ridicule something that they do not understand. By this scorn, they seek to diminish you, thus removing

the need to otherwise raise their own pitiful comprehensions. Kings have always had their fools about them, and we astronomers, too, have our mocking detractors.

There are two avenues of action open to you. The first is to show them how foolish they are being. Play up to their ridicule by dressing up in a clownish suit and lark about. In replacing the appearance of a superior activity for that of a circus clown, the thorn that pricks them is drawn. They will soon tire of watching out for you. You can then continue your observations unmolested. I have tried this with great success, and I can send you one of my cast-off clown suits if you require.

Secondly, you can turn the tables. You have, of course, a powerful surveillance instrument at your disposal. It should not take long for you to point it at all your detractors in turn, noting any individual peculiarities you see. Those individual examples of strangeness will be there, I assure you. Once a full complement of these idiosyncrasies has been accrued you have a toolbox with which to fight back.

For instance, as soon as a belligerent neighbor starts to heckle, point your instrument his way and bellow to the rest "Look at Mr. Thompson's pajamas. He's got teddy bears on them, ha ha" etc. Get the idea? Having the telescope, and the others too distant to disagree, they have no choice but to believe anything you say – so it doesn't even have to be true. Wicked, but nicely effective.

Some astronomers do indeed have the misfortune to have neighbors who do more than exhibit benign disinterest. They vocalize scorn, grow trees, and install deliberately misaligned security lights. Fortunately, as mentioned elsewhere, legislators are finally offering some protection against actions deemed as a "nuisance." While not aimed specifically towards the privations of astronomers, the legal recourse is at least at your disposal. In some instances this may be more painful than simply moving, but it may be cheaper.

(SOME ADVICE IS UNWELCOME.)

Dear Steve

I would like to complain about the advice you gave my husband concerning his problem. You may remember his failure to obtain detailed observations of our neighbor through her bathroom window due to obscuration by internal condensation. I feel that you should have told him to stop this perverse activity and not, as you did, tell him about the infrared interferometry technique commonly used for this purpose!

Renee
Edmonton, Alberta

Dear Renee,

Sorry, but it's the doctor-patient relationship thing. I am bound by the hypocratic oath to consider my client's interests first and foremost. I was bound to offer your spouse the sound technical advice he required to solve his problem – regardless of the situation created as a consequence of that solution. I trust that his investment in the equipment I recommended did the trick? Excellent.

However, your own letter places me in the invidious position of now being your advisor, not his. I am forced, therefore, by the same honorable compunction, to alleviate your distress by relaying information to you for the purpose of sabotaging his observations. This will make my therapist very happy. Do exactly as I tell you. Under my instructions, your husband will have purchased and fitted an R534 unit to the focus rack of his telescope. It is a black and gold cubic rectangle, with a little showerhead logo on the side. Found it? Good. At the top left hand corner of the back plate, you will find a small recessed grill held in by four screws at its corners. Remove these carefully using a cross-headed anti-static screwdriver. Careful! Lift the grill away slowly without touching the sides; a careless contact now with the rest of the chamber may set off the alarm. Can you see three thin wires just to the right of the steadily blinking red LED? There should be a white one, a blue one, and a red one. Snip the blue – and don't tug it or touch the other wires as you do so, for Christ's sake. Has the LED stopped blinking? If it has, good. If it hasn't, run as fast as you can; you only have 30 seconds. Are you still there? Wonderful. Replace the grill and screw down. Wipe a bit of dust on the screw heads so that they don't look disturbed. The equipment will no longer work. I do hope you're satisfied.

Of course, there is one thing left to do. I have included in my mailed reply to you another letter, addressed to your husband. It is the one which begins. . . "You wife has sabotaged your R534 unit. To repair, if you look at the top left hand corner of the back plate, you will find a small recessed grill held in by four screws at its corners. Remove these carefully using a cross-headed anti-static screwdriver. . . ."

To those who are now nervously measuring up their bathroom windows for blinds or curtains, in order to counter the possible presence of an indiscreet astronomer less than two blocks away, let me tell you something. With even a small telescope an astronomer can view the gravitational dance of binary star systems of blue and gold, watch the nativity of stars screaming birth pangs of fiery radiation that can be seen carving great chasms of empty space from their gaseous nursery, observe in wonder the coppery dimming of the full Moon as the interposing Earth's bulk robs it of sunlight during an eclipse, catch in a single glance the accumulated sunlight of 10 billion suns in the smudge of light that is a distant galaxy seen by the age of its light as it looked millions of years ago, then closer to home gaze upon the remains of cataclysmic impacts forged on the Moon by mountainous missiles during the violent creation of the Solar System. Even without a telescope, I can ponder that just beyond the thin atmospheric skin that my upturned gaze is piercing, our world is bathed in the infinitesimally weak afterglow of our universe's birth 14 billion years ago. Believe me, your bathroom holds no wonders that can compete with these things.

(PRESUMING THAT OUR MINDS ARE ON A HIGHER 'CELESTIAL' PLANE, THE GENERAL PUBLIC MAY BE FORGIVEN FOR BELIEVING THAT THE RUNNING OF ASTRO SOCIETIES PROCEEDS WITH SERENE DIGNITY AND INTELLECTUAL RIGOR. ALAS, THESE GROUPS ARE LITTLE DIFFERENT

THAN ANY OTHER ASSEMBLY OF HUMANS TRYING TO RUN THEIR OWN
AFFAIRS, I. E., SAVAGES.)

Dear Steve,

*The astro society to which I belong is in real trouble. Following a stormy annual
General Meeting this year, a dissident reactionary faction within our group was,
by devious means, able to capture a majority of the committee positions. This
coup d'état has resulted in deep divisions within our once harmonious group. In-
fighting is rife – expulsions of conservative members (on trumped up charges of
using a telescope in a public place) have occurred – with more threatened soon. It
has come to a head recently with the enemy phalanx forcing through measures
such as "membership is conditional on possession of a GO-TO scope," "academic
prowess to be denounced as elitism," "all lectures to be comprehensible to pre-
meiosis cytoplasm," and absurdly the "compulsory wearing of a silly hat by all
non-committee members." Lastly, they are inflicting a bizarre name change to
reflect the recent incorporation of the local Women's Guild of Needle Clackers
and Bobbin Bobbers. What can we do?*

Virgil
Reading, New York

Dear Virgil,

This problem is no stranger to astro societies all over the world (past and
present), including some that I have been a member of. This is why, in
1899, there was inaugurated an action group I belong to known merely by
the mysterious initials A.A.S. – American Astronomers Slain. (By a strange
quirk of fate, those initials were also used by another new organization for
astronomers in the same year.) We can be contacted by a protracted
process, the details of which have been sent to you in a plain brown
envelope. Charges run according to status; i.e., 'disappearance' of
ordinary members $25, active ordinary members not on the committee
$40 etc., eventually working up to committee officers $200 and chairman
$negotiable. We are good; really good. I am sure you can think of quite a
few prominent amateur astronomers who have mysteriously, suddenly,
ceased to be around. We did it. All of them. Those living in the UK have
access to our sister organization the B.A.A (British Astronomers Assassi-
nated). So go on. Make our day. . .

I hasten to stress that both the real AAS. (the American Astronomical Society) and BAA (the British Astronomical Association) are honorable organizations with published aims that do not include the removal of unpopular astronomers. So far as I know.

In theory, an astronomical society should be a group of benign amateur astronomers who occasionally give up the privilege of their largely solitary pastime to mix with like-minded enthusiasts for mutual comfort and reassurance. In reality, it can be a hot bed of personal and political intrigue that makes the murderous machinations of the medieval Borgias look like a preschool playgroup.

It almost goes without saying that a harmonious astronomical society is one that is well run. Those that are not soon succumb to acrimonious disintegration. It needs to be got right and kept that way. Having held various roles in the running of a successful society and observed from afar the explosive decay of others, may I offer a few fragments of wisdom. Organize yourselves within a mutually agreed framework of guidelines or rules; governing committees should always keep in mind that they serve other members and do not comprise an elite. Do not impose the same task on a single member for too long (particularly the position of chairperson). Above all, remember that everyone within the group has joined for a single purpose – to enjoy astronomy. *Forget the last at your peril.*

Fig 8.6 [Image by the author]

(IT HAS BEEN AN OBSERVATION [HERE AND ELSEWHERE] THAT LOCAL ASTRONOMICAL SOCIETIES ARE OVER-BLESSED WITH LONELY MEN. I DO NOT ENTIRELY UNDERSTAND THIS, BUT IT MAY HAVE SOMETHING TO DO WITH THE FACT THAT OBSERVING THE UNIVERSE CAN BE, DARE I SAY, AN ANTISOCIAL PAST-TIME; ONE MAN, ONE TELESCOPE, ONE SKY. SOME GUYS JUST DON'T GET OUT ENOUGH TO OVERCOME THEIR SHYNESS!)

Dear Steve,

I am a rather quiet sort of man who never goes out much. I am very shy and retiring and terribly nervous of any sort of contact with.. ..er.. .gir. . .er. . .wom. . .er.. people of the opposite ssss. . .er..mmmm, let's put it this way, I'm pretty lonely. The only spark of life that I have is reserved for my fantastic astronomy. I am out all night in my backyard, casting my gaze across the universe, drinking in the wonders. A blue sky deepening to indigo is the best turn-on I can experience. That euphoric feeling when I think of those potentially spectacular sights gives me a tangible tingling of excitement throughout my whole body. It's like, wow! The sky is a panorama of discovery for me. It is a comfortable territory that holds no secrets or threats for me, yet still offers that tantalizing possibility of a new sighting of the unknown. But, I am on my own. I have no partner with which to share this heaven.

Apart from the trouble with my bashfulness, the stutter, and the fact that I fall over my feet a lot, I find that as soon as I start talking about astronomy (the only thing I can really talk about) any young la..la. . .lady soon begins to yawn. They try to stifle their yawns to be polite, but I can tell. The only good thing to come out of such conversations is that I'm getting quite adept at dealing with dislocated jaws. But then tears stream down their cheeks and soon after they become unconscious. I guess I am the archetypal Nerd. But I still have my natural urges, and I have the right to express them, don't I? These cannot be sated by astronomy alone. Can you suggest how I might meet the g..g..gir..partner of my dreams? I don't mind if she yawns too much – so long as she stays awake for the important bits.

Jed
New Haven, Connecticut

Dear Jed,

You'll be glad to hear there is help at hand. You have obviously not heard of **NOVA** (**N**erdish **O**rganization for **V**acuous **A**stronomers) of which I am patron. I'm sending you details by separate post, but in the meantime I will give you a brief lowdown for the sake of other sad astronomers who are no doubt out there. NOVA offers a new life through a dynamically laid back social network that, for the benefit of everyone else, brings together mindless monological morons like you to a warm and comforting risk-free environment of totally stultifying turgitude. This is so we at least know where you are. Be in absolutely no doubt that there are plenty of girls who are just as single mindedly tramlined as yourself. Within NOVA you get a catalog and can take your pick. We sort potential friends by such things as height, hair color, financial security, weight, and personality disorder. We don't list their interests, for the obvious reason that they have none, bar one. But be warned. One couple recently was so numbingly matched in boredom that neither had the will to break off a mutually protracted stare. There was little to do but watch them slowly starve to death!

If someone, anyone, can supply me with a reasonable explanation as to why there is a dearth of women attending amateur astronomy groups can they please let me know! The history of astronomy is strewn with women who have indulged in and made serious contributions to the science of astronomy. When I talk to most women about the subject (those I can catch) they obviously have as much passing interest in the subject as any man. Some are even enthusiastic about it. Female professional astronomers are obviously just as sharp and energized as their opposite gender colleagues. Why, then, are there so few of the gentler sex attending local astronomical groups? I'll readily admit that their representation is much better now than in past years, but still not good enough – and far less than their proportion of the population. I tend to do an occasional head count of my own local group and those I visit – and it's never better than 10%. Something isn't being done right, and I'm fascinated to know what that is. If you know, ladies, please call!

It may be that with a pattern in the past of male domination of these groups, we have a bit of historical momentum to divest first. Perhaps facing the overwhelming number of resident gentlemen at a meeting is daunting. I sympathize with that concern, especially as I know only too well that the unexpected attendance of a female amateur astronomer elicits interest on many fronts. I did, after all, marry one!

(THE MARCH OF TECHNOLOGY IS INCESSANT. MEDIEVAL FOLK WOULD TODAY NOT BE ABLE TO EVEN BEGIN GUESSING AT THE PURPOSE OF MOST OF OUR EVERYDAY IMPLEMENTS. BUT EVEN A CONTEMPORARY HUMAN MAY HAVE PROBLEMS IN DECIPHERING WHAT SOME THINGS DO.)

From: *Shepherd Jail[cage@shepdpolice.com]*
To: *Dear Steve[doctorsteve@help.com]*
Subject: *Ref. 2005231/AB-3*

Hi Steve,

My name is Laura-Li. You don't know me (although I'm sure you'd like to). I hope you don't mind my contacting you out of the blue like this, but I had a choice between making a telephone call or sending an email. So I have been given permission by this nice Kentucky police officer to contact you directly by email. He only let me use his terminal if I'd say he was handsome – which of course he is!

All I was doing was trying out my new GPS celestial finder. It's the 'Star-Guide Deluxe' made by Clouds-B-Damned. Do you know the one? It's one of those devices that you hold in your outstretched hand and point at the sky and it tells you all about the celestial object it's pointed at. Well, there I was about 11 p.m. tonight, just minding my business in the dark walking slowly around pointing this thing in all directions to test it out when Wham!, three police cars suddenly appeared with sirens going and lights flashing. I thought they were racing off to an emergency, but they closed in where I was standing and skidded to a halt around me, dust flying everywhere. I was so scared by the sudden noise. I nearly crapped myself when the cops flung open the car doors and dropped to the ground in a circle around me with their hand-guns drawn! One of them, the sergeant I think, yelled at me to lower my arm and drop my weapon. When I shouted back that I didn't have a weapon, they fired a warning shot at me that struck the ground just inches from my new red high-heeled shoes! So, I then figured that I'd better do as I was asked and explain later. So I dropped my expensive sky-pointer. Would you believe it! The damn thing smashed on the sidewalk into a zillion pieces! They rushed me and threw me to the ground, then hand-cuffed my hands behind my back. Boy, did that hurt. And being a warm evening (no bra, lovely) I am only wearing a short skirt tonight so the tussle put some runs in my high-gloss pantyhose, too, damn it!

They arrested me, there and then, for carrying an offensive weapon, resisting arrest, and prostitution! I tried to explain what I was doing, but my device was so broken that I couldn't prove to them what it was. They just said it was obviously a home-made firearm of some sort (especially when I made the mistake of saying it had been fitted with a laser pointer). They just added that it was lucky that a concerned member of the public had phoned them in time before I had shot someone. My God! Who do they think I am, Annie Oakley? I didn't have my handbag on me so couldn't show them my astro club membership card.

So, Steve, could you please reply to this email urgently and tell them I am an innocent astronomer who was simply using a star-pointing device that loads of us star watchers have. I have tried explaining, but they just won't believe me. They don't accept that someone would go out into the dark late at night and point around at the sky for amusement. I mean, you meet so many friendly people this way!

In Kentucky, we have a thriving amateur astronomical community and lots of us girls have been using these things. It's just awful. They currently think me so dangerous that they won't even consider granting me bail. They put me in this pen where they've got all the other girls who've said they're doing astronomy, too. I mean, just how much evidence do they need? It's a disgrace! Help, please, soon as you can. If you can help me and the other girls, I'm sure we would all be so very grateful. My case reference is 2005231/AB-3. Thanks.

From: Dear Steve[doctorsteve@help.com]
To: Shepherd Jail[cage@shepdpolice.com]
Subject: Ref. 2005231/AB-3

Hi Laura-Li!

What an interesting life you lead. Sorry for my delay in responding to your clearly urgent request for assistance. I was out observing until very late last night, and I have only just woken up and finished my coffee. I do hope you're overnight ordeal wasn't too awful. I have to say, what an imaginative combination of two worthwhile careers. Of course I'll be happy to vouch for you. Tell the officer that I, too, indulge in this activity (though not of course in high heels). Tell him to ring me on the number attached, and between us I am sure we will come to some arrangement. Maybe, when you've got yourself a replacement for your Star-Guide Deluxe, we could compare devices? Call me.

It was inevitable that with the commonplace development of GPS and acceler-ometers, these technologies would be combined with astronomy databases into devices that not only know when and where they are but know exactly which way they are pointing. Thus are created today's handheld guides to the night sky. They can not only inform the user of the target's identity but also spew chapter and verse on its characteristics. From personal experience I have to add that this noble ambition has resulted in perhaps marginally less than 100% accuracy, but at least their electrically humming heart is in the right place.

However, walking around in the dark, wistfully aiming an unidentifiable object in random directions (at least so far as a distant observer can determine), obviously embodies elements of a threatening manner. There have been anecdotal indications (at least as related on Internet blogs) that law enforcement officers have indeed been called to intercept groups wielding what appear to be firearms! Whether these are actual events or reports mischievously posted by rueful competitors it is difficult to tell. But having used the devices I find these plausible occurrences. One can just imagine the scene. A black & white screeches to a dust-skidding halt before confused and headlight-blinded astronomers. "Drop your weapons" demand the gun-toting officers. "But these aren't weapons" plead the alarmed astronomers. "Likely story" comes the reply, as all are herded away to the sergeant's desk downtown.

Nevertheless, there is a moral to this tale, and it is this. If you are arranging an observing session – especially in an area whose inhabitants are unused to such nocturnal activity, it is wise – nay essential – that the local police are informed. My own group in the UK often venture into the local countryside seeking dark sites, but they assiduously advise regional law enforcement when they do so. Occasions when this measure has been overlooked have often resulted in a "visit" by local constables, although the benign result usually has them looking through our telescopes, rather than making arrests.

(I AM DISTURBED NOT A LITTLE THAT MANY OF THE DEAR STEVES CONCERN ENVY. A TRUE REFLECTION, I FEAR, THAT IT'S NOT SIMPLY PRESENT – IT'S RAMPANT.)

Dear Steve,

I have good reason to think that my wife is having an affair with another astronomer. I am sure that the reason for this is because he has a larger telescope than I have. As a result, when it gets dark, he has more to show her than I do. I think I may lose her, because I simply cannot afford to get bigger one. What can I do?

Luther
Las Vegas, Nevada

Dear Luther,

I believe the problem is not one of size but of frequency. I infer, since you make no comment on this yourself, that because yours is smaller, you use it less often than you would otherwise. Possibly it is this that has driven your wife to seek an evening's satisfaction with someone whose telescope exceeds yours. Getting yours out more frequently and spending more time in the garden with it will soon make your wife realize how lucky she is. It may help if you give her a little one to play about with while you are not there, so she can get the better feel of the subject. Take comfort in the thought that it is likely she is not enjoying her clandestine meetings with this 'other' astronomer, as larger instruments are always far more difficult to maneuver.

Comparing telescopes used to be easy. Reflector or refractor, inch for inch, the bigger aperture was likely to be better. No so now. The introduction of erotic (oops, sorry) exotic glasses into telescope design, such as the use of fluorite crystal, has created telescopes that are far better corrected for the various aberrations previously inherent in most refracting telescopes. Such instruments, termed apochromatic refractors (APOs), may be physically smaller in aperture than conventional achromatic refractors, but by virtue of their enhanced contrast and color correction are able to show far more detail.

Having said that, the old adage about "it's not what tools you have it's the way that you use them" very much applies to telescope use. Much of the processing of what you see through a telescope happens behind the eye, not in front of it. Practiced patient observation will make visible through the same telescope detail not seen with an inexperienced eyeball. I come across this many times, quite often during public observation evenings when those not used to looking through optical instruments fail to see detail pointed out to them by long-standing observers. This is why I always warn new telescope owners not to expect the best from their telescope right away. This is through no fault of the instrument. It has to wait for the human side of the optical partnership to catch up!

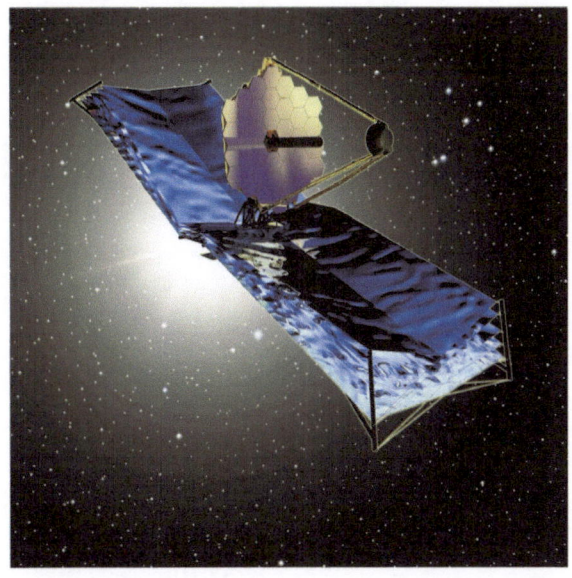

Fig 8.7 Hubble's infrared successor. The 6.5-m diameter gold-coated beryllium James Webb space telescope. In the end, regardless of size, it's how you use a tool that counts [Image courtesy of NASA.images.org. Marshall Space Flight Center Collection]

Index

A

Accelerometers, 212
Accessories Anonymous, 92–93
Achromatic, 65, 214
Adams, Douglas, 18
Affair, 50, 55, 213
Alcock, George, 22
Alcohol, 24, 25, 110–111, 122, 139, 180, 195
Almagest, 30
Alphonsus, 163
Aluminizing, 51
American Astronomical Society, 126, 142, 196, 208
Andromeda Galaxy, 185
Annual General Meeting, 207
Antique, 14, 54, 77
Ants, 57–58
Aperture, 24, 36, 40–41, 63, 72, 113, 117–118, 174, 214
Aperture Fever, 41
Apochromatic, 214
Apollo, 20, 54–55, 102, 163
Arc-lights, 130
Armageddon, 137
Asimov, Isaac, 133
Asteroids, 129, 137, 178, 181
Astigmatism, 69, 79, 88
Astroblock, 96
Astrology, 11–12
Astronaut, 54–55, 70–72, 102, 104
Astronomical authors, 170
Astronomical ephemeris, 169
Astronomical societies, 46–47, 49, 73, 87, 113, 115, 126, 142, 181, 195–196, 198–199, 208–209
Astronomy quizzes, 10
ATM, 63
Averted vision, 147

B

Backlash, 35, 46
Barlow, 88
Barnard, Edward E., 150
Bayer, Johann, 28
BBC, 168
Beagle, HMS, 85
Beatles, 13
Bell Burnell, Jocelyn, 198
Bicycle, 175
Binoculars, 22, 60–61, 114–115
Blaaaaa, 166
Black Light Bulb Brigade, 141
B.O., 191
Borgia family, 208
BPM 37093, 13
Bradbury, Ray, 150
Bradfield, Bill, 129
Braine, David, 158
Brewing, 110
Brewster, Lady Jane, 183
British Association, 183
British Astronomical Association, 126, 142, 196, 208
Burroughs, Edgar Rice, 150

S. Ringwood, *Astronomers Anonymous*, DOI 10.1007/978-1-4419-5817-4,
© Springer Science+Business Media, LLC 2010